Coding Dimensions and the Power of Finite Element, Volume, and Difference Methods

Abdulsattar Abdullah Hamad
University of Samarra, Iraq

Sudan Jha
Kathmandu University, Nepal

A volume in the Advances in Systems Analysis, Software Engineering, and High Performance Computing (ASASEHPC) Book Series

Published in the United States of America by
IGI Global
Engineering Science Reference (an imprint of IGI Global)
701 E. Chocolate Avenue
Hershey PA, USA 17033
Tel: 717-533-8845
Fax: 717-533-8661
E-mail: cust@igi-global.com
Web site: http://www.igi-global.com

Library of Congress Cataloging-in-Publication Data

CIP DATA PROCESSING

2024 Engineering Science Reference
ISBN(hc)- 9798369339640 | ISBN(sc)- 9798369349014 | eISBN- 9798369339657

British Cataloguing in Publication Data
A Cataloguing in Publication record for this book is available from the British Library.

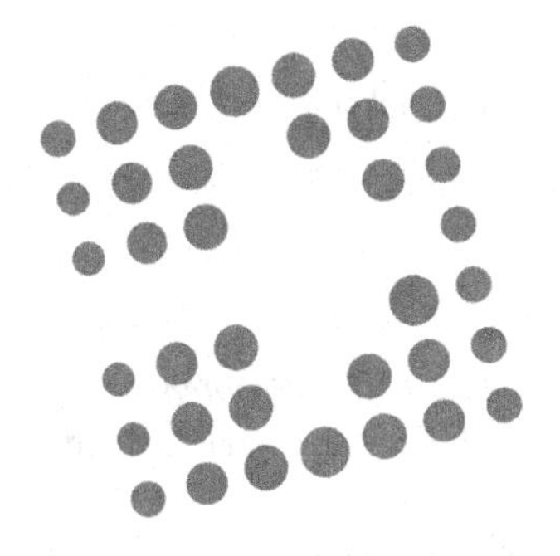

Advances in Systems Analysis, Software Engineering, and High Performance Computing (ASASEHPC) Book Series

MISSION

Vijayan Sugumaran
Oakland University, Rochester, USA

ISSN:2327-3453
EISSN:2327-3461

The theory and practice of computing applications and distributed systems has emerged as one of the key areas of research driving innovations in business, engineering, and science. The fields of software engineering, systems analysis, and high performance computing offer a wide range of applications and solutions in solving computational problems for any modern organization.

The **Advances in Systems Analysis, Software Engineering, and High Performance Computing (ASASEHPC) Book Series** brings together research in the areas of distributed computing, systems and software engineering, high performance computing, and service science. This collection of publications is useful for academics, researchers, and practitioners seeking the latest practices and knowledge in this field.

Coverage

- Metadata and Semantic Web
- Network Management
- Storage Systems
- Virtual Data Systems
-

IGI Global is currently accepting manuscripts for publication within this series. To submit a proposal for a volume in this series, please contact our Acquisition Editors at Acquisitions@igi-global.com or visit: http://www.igi-global.com/publish/.

The (ISSN) is published by IGI Global, 701 E. Chocolate Avenue, Hershey, PA 17033-1240, USA, www.igi-global.com. This series is composed of titles available for purchase individually; each title is edited to be contextually exclusive from any other title within the series. For pricing and ordering information please visit http://www.igi-global.com/book-series/advances-systems-analysis-software-engineering/73689. Postmaster: Send all address changes to above address.

Titles in this Series

For a list of additional titles in this series, please visit: www.igi-global.com/book-series

•

Revolutionizing Curricula Through Computational Thinking, Logic, and Problem Solving
Mathias Mbu Fonkam (Penn State University, USA) and Narasimha Rao Vajjhala (University of New York, Tirana, Albania)
Engineering Science Reference • copyright 2024 • 245pp • H/C (ISBN: 9798369319741) • US $245.00 (our price)

Harnessing High-Performance Computing and AI for Environmental Sustainability
Arshi Naim (King Khalid University, Saudi Arabia)
Engineering Science Reference • copyright 2024 • 401pp • H/C (ISBN: 9798369317945) • US $315.00 (our price)

Recent Trends and Future Direction for Data Analytics
Aparna Kumari (Nirma University, Ahmedabad, India)
Engineering Science Reference • copyright 2024 • 350pp • H/C (ISBN: 9798369336090) • US $345.00 (our price)

Advancing Software Engineering Through AI, Federated Learning, and Large Language Models
Avinash Kumar Sharma (Sharda University, India) Nitin Chanderwal (University of Cincinnati, USA) Amarjeet Prajapati (Jaypee Institute of Information Technology, India) Pancham Singh (Ajay Kumar Garg Engineering College, Ghaziabad, India) and Mrignainy Kansal (Ajay Kumar Garg Engineering College, Ghaziabad, India)
Engineering Science Reference • copyright 2024 • 354pp • H/C (ISBN: 9798369335024) • US $355.00 (our price)

701 East Chocolate Avenue, Hershey, PA 17033, USA
Tel: 717-533-8845 x100 • Fax: 717-533-8661
E-Mail: cust@igi-global.com • www.igi-global.com

Table of Contents

Preface

In today's ever-evolving technological landscape, the ability to effectively utilize numerical methods in solving complex engineering problems has become a crucial skill. Whether you are a seasoned engineer or a curious learner, *Coding Dimensions: Unleashing the Power of Finite Element, Volume, and Difference Methods* equips you with the knowledge and skills needed to harness the power of these indispensable techniques. This book bridges the gap between theory and practice, providing hands-on experience in implementing these methods through coding examples and real-world applications, particularly in the simulation of dynamic systems and the resolution of partial and ordinary differential equations.

This volume aims to empower engineers, researchers, and students to push the boundaries of innovation by offering a solid foundation in numerical methods and their practical implementations. With a particular emphasis on hands-on coding examples, "Coding Dimensions" serves as a comprehensive guide to tackling engineering challenges using finite element, volume, and difference methods. By delving into the potential of these techniques, readers are invited to unlock their power through coding and embark on a journey of computational exploration.

The target audience of this book includes professionals and researchers working in the fields of education, scientific research, computer science, and information technology. Moreover, this book provides invaluable insights and support for executives and researchers seeking to combine mathematics with other sciences, thereby keeping pace with technological advancements. By doing so, it empowers engineers, researchers, and students to innovate and excel in their respective fields through a thorough understanding of numerical methods and their applications.

Covering a wide array of topics, this book starts with the fundamentals of numerical methods, introducing their applications in engineering and offering an overview of finite element, volume, and difference methods. It delves into the mathematical foundations and principles behind these methods, essential for numerical analysis and error estimation. The book also addresses the generation of structured and unstructured meshes, techniques for mesh refinement, and methods for handling irregular and complex geometries, ensuring accuracy and efficiency in computations.

Readers will find comprehensive discussions on solving linear and nonlinear problems, with chapters dedicated to linear algebraic systems, direct and iterative matrix-solving methods, preconditioning, and parallel computation techniques. The treatment of nonlinear problems includes solutions for nonlinear equations, Newton-Raphson methods, and applications in structural, thermal, and fluid analysis. For time-dependent problems, the book explores time integration schemes, stability and accuracy analysis, and solutions for heat conduction, wave propagation, and diffusion problems using fully implicit and explicit schemes.

The book further delves into parallel computing techniques, covering distributed memory architectures, domain decomposition, message-passing interfaces, and GPU utilization for numerical computations. Advanced topics include higher-order basis functions, adaptive mesh refinement, design optimization, multiphysics simulations, and coupling methodologies. Additionally, the book provides insights into software development and best practices, introducing programming languages used in numerical methods, software engineering principles, verification and validation of numerical models, and relevant software libraries and open-source projects.

We hope that *Coding Dimensions: Unleashing the Power of Finite Element, Volume, and Difference Methods* will serve as a vital resource for those dedicated to mastering numerical methods. May it inspire and guide you as you explore the vast possibilities within computational engineering.

Chapter 1: Agile Practices for Software Development for Numerical Computing

B. Tamilarasan, Venkata Surya Bhavana Gollavilli, Poovendran Alagarsundaram, BalaAnand Muthu

Software development is inherently complex, involving numerous variables that require careful management. Traditional Software Engineering introduced extensive documentation and artifacts to impose order on this process, but over time, the bureaucratic nature of these methodologies became a hindrance. Agile Techniques emerged as a solution, optimizing time and fostering better team dynamics and customer satisfaction. This chapter explores how Agile Practices have revolutionized software development for numerical computing by enhancing communication, trust, and productivity through simplicity and adaptability. The authors delve into how these techniques have gained widespread acceptance and effectiveness in modern software development environments.

Chapter 2: Building Blocks – Meshing

Harikumar Nagarajan, B. Tamilarasan, Thanjaivadivel M

This chapter focuses on the analysis of structural building blocks subjected to finite displacements using incremental sequential techniques. Employing the Newton-Raphson iteration approach for various load stages, including scenarios where the global stiffness matrix's determinant approaches zero, the authors utilize advanced Fortran programming for structural analysis. Verification is performed using the Structural Analysis Program (SAP), and multiple case studies demonstrate the software's application. The chapter compares axial stiffness of different elements and the structural responses modeled through stiffening, snap-through, and softening load-displacement curves, providing a comprehensive understanding of meshing techniques and their practical implications.

Chapter 3: Fundamentals of Numerical Methods

Shihab A. Shawkat, M. Thivagar

This chapter provides an extensive overview of numerical methods, their fundamental concepts, and their application in solving practical engineering problems. It highlights the importance of numerical methods in understanding and addressing differential equations critical in engineering. The authors cover the potential and limitations of these methods, offering insights into the finite element method's application across various engineering challenges. This exploration sets the stage for a deeper appreciation of how numerical methods enhance our comprehension of complex phenomena in the engineering domain.

Chapter 4: Handling Irregular and Complex Geometries

C.B. Sivaparthipan, Waleed Khalid Al-Azzawi

Addressing the challenges of analyzing structural components under multiple loading conditions using the finite element method (FEM), this chapter introduces a practical method for obtaining optimal meshes in linear static refinement problems. The proposed approach enables the use of a single mesh for various load cases, streamlining the analysis process and achieving the desired error margins. The authors validate their method through two-dimensional examples, demonstrating its efficacy in handling irregular and complex geometries efficiently.

Chapter 5: Mathematical Foundations and Principles Behind These Methods

Noor Kadhim, Abdulsattar Hamad

This chapter delves into the historical context and essential definitions of numerical methods in engineering, emphasizing accuracy, precision, convergence, and stability. It explores different types of errors, their quantification, and the use of Taylor polynomials for approximating mathematical functions. By establishing a strong mathematical foundation, the authors provide a comprehensive understanding of numerical analysis, setting the stage for more advanced discussions on numerical methods and their applications.

Chapter 6: Parallel Computing Techniques

Alnoman Tayyeh, Akram Shather, Saja Anaz, Firas Jasim

Focusing on meshless methods for solving partial differential equations, this chapter highlights the advantages and challenges of these techniques, particularly in scenarios requiring frequent remeshing. The authors discuss the application of parallel programming techniques, specifically "Open Multi-Processing," to accelerate computational codes for electromagnetic models based on meshless methods. The chapter provides a detailed analysis of parallel computing's role in enhancing the efficiency of numerical computations in complex engineering problems.

Chapter 7: Software Development and Best Practices

Ahmed Ibrahim Turki, Sushma Allur, Durga Praveen Deevi, Punitha Palanisamy

This chapter examines the evolution of software programs capable of solving complex problems quickly, focusing on non-linear equations and numerical integration. The authors compare the effectiveness of the Secant and Newton-Raphson methods in Python, alongside Simpson's methods for numerical integration using Pascal-based programs. Through detailed programming tests and data analysis, they demonstrate the accuracy and efficiency of these numerical methods, offering insights into best practices for software development in numerical computing.

Chapter 8: Solving Linear Problems

Alnoman Tayyeh, Akram Shather, Firas Jasim, Saja Anaz

Exploring the dynamics of heat exchange in industrial processes, this chapter presents a study of flow in a packed bed through computer simulation. The authors consider a cylindrical bed made up of spheres, using Gambit software for geometry and mesh generation and Fluent software for flow simulation. Simplifications such as flow periodicity and symmetry are employed to make the simulation feasible. The chapter provides detailed insights into the simulation process and its application in structural, thermal, and fluid analysis.

Chapter 9: Solving Nonlinear Problems

Akram Shather, Mohanad Shadhar, Harish Bhandari

In the context of technological advancements and Industry 4.0, this chapter discusses the evolving role of engineers in product development. It emphasizes the importance of CAD/CAE tools in creating realistic models and integrating various numerical and computational resources. The authors highlight the need for continuous learning and adaptability in the face of digital manufacturing trends, such as Augmented Reality, 3D Printing, and Virtual Reality. This chapter underscores the educational sector's shift towards "School 4.0," preparing professionals for the dynamic demands of modern industry.

Chapter 10: Time-Dependent Problems

Riad Al-Hamido

Introducing new types of temporal open and closed sets in temporal topological spaces, this chapter explores their relationships and properties compared to traditional topological spaces. The authors present original theorems and remarks, providing a groundbreaking study in the field of temporal topological spaces. This chapter offers innovative insights into the mathematical modeling of time-dependent problems, expanding the theoretical framework for future research.

Chapter 11: Utilizing Graphics Processing Units (GPUs) for Numerical Computations

Alnoman Tayyeh, Akram Shather, Husam Hussein, Luma Abdalbaqi

Tracing the evolution of GPUs from fixed-function rendering devices to versatile parallel processors, this chapter discusses their growing role in numerical computations. The authors explore the integration of GPU and CPU cores within single chips and the use of GPUs in supercomputers. By examining the application of GPUs in accelerating numerical methods, this chapter highlights the significant performance improvements achieved in computational engineering.

Chapter 12: Validation and Verification of Numerical Models

Bahaa Hayder Mohammed, Yas Khudhair Abbas, Basava Ramanjaneyulu Gudivaka, Sri Harsha Grandhi

Focusing on the critical concepts of validation and verification in numerical modeling, this chapter aims to validate a finite element method-based code used for simulating incompressible fluid flows. The authors conduct numerical simulations and experimental tests, comparing results with literature data. While discrepancies are noted, the pressure coefficient curves obtained from simulations align reasonably well with existing data. This chapter emphasizes the importance of rigorous validation and verification processes in ensuring the reliability of numerical models.

As editors of *Coding Dimensions: Unleashing the Power of Finite Element, Volume, and Difference Methods*, we are delighted to present this comprehensive guide to numerical methods in engineering. This volume brings together a wealth of knowledge and expertise from contributors across the globe, each offering unique insights into the practical applications and theoretical foundations of these indispensable techniques.

The chapters in this book meticulously cover the spectrum of numerical methods, beginning with an exploration of Agile practices tailored for software development in numerical computing. This sets the stage for understanding the dynamic nature of software engineering and the pivotal role of Agile techniques in enhancing efficiency and team dynamics.

We then transition into the technical core of numerical methods, starting with the fundamental principles and moving through intricate discussions on meshing techniques, handling complex geometries, and the mathematical foundations essential for accuracy and precision in computations. Each chapter builds on the previous, providing a seamless progression from basic concepts to advanced applications.

Particular emphasis is placed on solving both linear and nonlinear problems, with detailed methodologies and case studies demonstrating the practical implementation of these techniques. The inclusion of parallel computing and GPU utilization chapters highlights the cutting-edge technologies driving modern computational methods, ensuring that readers are equipped with the knowledge to leverage high-performance computing resources effectively.

The chapters on software development best practices and programming languages serve as invaluable resources for practitioners looking to enhance their coding proficiency and apply numerical methods to real-world problems efficiently. These discussions are supplemented with practical examples and comparative analyses, making complex concepts accessible and actionable.

In addressing time-dependent problems and the novel exploration of temporal topological spaces, this book pushes the boundaries of traditional numerical methods, introducing new theoretical frameworks and practical insights. These chapters underscore the innovative spirit that drives computational engineering and opens new avenues for research and application.

Finally, the importance of validation and verification of numerical models is underscored, ensuring that the methods and techniques discussed throughout the book are grounded in rigorous testing and real-world applicability. This emphasis on accuracy and reliability is crucial for advancing the field and fostering trust in computational solutions.

We believe that this book will serve as an essential reference for engineers, researchers, and students alike. It not only provides a deep dive into the theoretical aspects of numerical methods but also bridges the gap to practical implementation, encouraging a hands-on approach to learning and discovery.

In an era where technology and engineering are rapidly evolving, this book is a testament to the collaborative effort and dedication of its contributors. It is our hope that readers will find inspiration and guidance within these pages, enabling them to tackle the complex challenges of modern engineering with confidence and creativity.

We extend our heartfelt thanks to the authors for their invaluable contributions and to the readers for embarking on this journey of computational exploration with us. May "Coding Dimensions: Unleashing the Power of Finite Element, Volume, and Difference Methods" empower you to innovate and excel in your endeavors.

Abdulsattar Hamad
University of Samarra, Iraq

Sudan Jha
Kathmandu University, Nepal

Acknowledgement

I would like to express my deep thanks and deep gratitude to everyone who contributed and supported me during the journey of writing this work *Coding Dimensions and the Power of Finite Element, Volume, and Difference Methods*. Firstly, I would like to thank Dr. Sudan Jha for his endless support and constant motivation. I also want to thank the IGI Global publishing team who worked hard to bring this book into existence.

I would also like to mention the friends and colleagues who have provided me with emotional support and advice during this journey. Without you, this beautiful book would not have seen the light.

Finally, my sincere thanks go out to every reader who now owns this book. I hope it inspires and benefits you all.

With all love and gratitude,

Dr. Abdulsattar Abdullah Hamad

Dr. Sudan Jha

Chapter 1
Agile Practices for Software Development for Numerical Computing

B. Tamilarasan
Madurai Kamaraj University, India

Venkata Surya Bhavana Harish Gollavilli
Under Armour, USA

Poovendran Alagarsundaram
Humetis Technologies Inc., USA

BalaAnand Muthu
Tagore Institute of Engineering and Technology, India

ABSTRACT

Software development is an extremely complex activity. It brings with it a very large number of variables. Software engineering was created to try to bring order to this activity. It initially came with the engineering bureaucracy and brought a very large diversity of documents and artifacts. Over time, the excessive amount of bureaucracy within the existing methodologies in software engineering became a burden in certain projects and there was a need to create agile techniques so that time is optimized and better applied, also bringing new ways of managing the team and developing the software. These new techniques have brought a great evolution in the exchange of experience, communication, transmission of knowledge, people's trust, customer trust. This increases the team's productivity and also increases customer satisfaction. Agile techniques have as their basic principles simplicity and easy adaptation to changes, making them very well accepted by systems development

DOI: 10.4018/979-8-3693-3964-0.ch001

teams and gaining more and more space among them in the current scenario.

INTRODUCTION

Agile methodology is a model and philosophy that proposes alternatives to traditional project management and improves the process of developing a product or service. The objective is to make deliveries with Agility (faster and more frequent) as the customer's needs arise.

Nowadays, every technology company or company that develops systems uses agile practices. Or at least they believe so. Whether you're new to software development or started decades ago, your work today is influenced by agile methods.

The agile method was formally launched in 2001 when technologists wrote the Agile Manifesto. The idea was to look for ways to be more efficient when programming/coding and help companies stand out in the software development market.

Although the agile methodology originated in technology, it has reached many other business areas due to its effectiveness. To successfully implement Agile in places where it is rarely used, organizations should begin with comprehensive Agile training and education under the direction of knowledgeable coaches. Small experimental initiatives and culturally appropriate techniques can show results and foster trust. Sustaining the change requires strong leadership support, adaptation, and ongoing feedback. The agile methodology proposes avoiding extensive and time-consuming product launches in favor of smaller, more incremental objectives.

In this type of development, teams continually evaluate their tasks and processes to improve their products and performance. Hardware selection, caching, indexing, partitioning, sharding or replication, and schema/query optimization are some aspects that guarantee a database model's scalability and performance. These procedures make it possible to handle increasing data quantities and user loads efficiently, preserving performance and dependability.

Today, there are many different types of tools, such as Scrum, Lean, and Extreme Programming (XP), but they all share principles of the broader agile approach. Check out this Accelerator article on five agile practices to improve your software development area and thus make it more effective.

Due to the magnitude and academic challenge that this project implies, the members of the Research Seedbeds showed concern because, until now, the developments had been prototypes created without formal documentation and, even worse, without the methodological rigor that software development required. There were knowledge gaps and uneven procedures in previous software development because formal documentation and structured methods were lacking. The ambiguous criteria made debugging more difficult, delayed growth, and impeded onboarding.

Errors during updates due to dependency management concerns resulted in delays and increased expenses. This recognition exercise raises the need to apply software engineering models that allow for managing development processes and making this project scalable and sustainable over time.

The need for a methodological proposal supporting this project and the other software processes that the Internet of Things hotbed carries out is recognized, given its emphasis on developing software applications for mobile devices. It begins, then, with tracking the methodologies that can be applied to software projects for mobile applications, and in this sense, (Al-Ratrout et al., 2019)

states that traditional software engineering techniques can be applied to the development of mobile applications; however, it mentions the relevance of using Scrum and other agile methodologies for this type of project.

The little recognition that the incubator members have about these agile methodologies demands investigating a framework for agile practices, which fosters a culture of adaptability to changing requirements and continuous feedback in developing a software project. Furthermore, they put people above processes, something completely different from traditional methodologies. These values aroused the students' interest, and they saw the opportunity to investigate the agile practices that can be applied in research hotbeds.

SOFTWARE PROCESS MODELS

Currently, companies seek to improve their processes to achieve higher quality in their products in the shortest possible time, reducing costs and minimizing errors in manufacturing their products to keep customers satisfied. Software development companies are no exception; every day, they strive to be more agile in all their processes to obtain high-quality software that meets customer needs. Businesses use test automation, CI/CD, Agile, DevOps, parallel testing, shift-left testing, risk-based testing, and feedback loop monitoring to balance speed and quality in software development. These procedures enable quick development, early problem identification, effective deployment, and ongoing improvement, guaranteeing the timely delivery of high-quality software.

For a software product, whether a web application or mobile application, to maintain quality criteria, in addition to satisfying customer expectations, it must have the following characteristics (Ayada and Hammad, 2023):

- Functionality: The product must fulfill the customer's requirements.
- Reliability: The system must be created to avoid failures, and when they occur, it can be recovered.
- Usability: A software system must have a suitable user interface for customers.

° Efficiency: The product must use all its resources and not waste them.

The problem with process improvement in software engineering is that the basis of development is the requirements specified by the Client at the beginning of the project, which makes the result unpredictable. Many times, functionally, the systems do not fulfill their purpose. There have been years of delays in the delivery of the systems, the costs exceed what was budgeted, and the systems have very low performance.

These poor results are not due to the developers' inexperience or lack of knowledge; they are just due to a lack of software engineering techniques or skills necessary for a large-scale project.

The software industry has responded by applying engineering to software creation, creating different methodologies, models, and practices that structure the activities required to develop software systems, and increasing the quality of the product.

In the 1970s, Winston Royce created the so-called waterfall model. This model served as the basis for the formulation of structured analysis, which was one of the precursors to the application of standardized practices within software engineering.

Over time, certain disadvantages became evident in the waterfall model since the product's results were seen at the end of the process, which required the user to be patient. This gave way to the creation of other models, such as the spiral model and the evolutionary model, where the objective is to build an initial version that is exposed to the Client and refined with their comments and suggestions.

These models gave way to the creation of different methodologies called traditional methodologies, from which processes have been deployed, such as the Rational Unified Process, which incorporates the stages of the spiral model cycle.

Better known as RUP (Rational Unified Process), it is one of the most influential processes in the software industry, and it was developed and supported by Rational® Software from IBM. As previously explained, the RUP life cycle is based on the spiral model that organizes tasks into phases and iterations (Shafiee et al., 2020).

RUP divides the process into four phases, within which several iterations are carried out in varying numbers depending on the project, and more or less emphasis is placed on the different activities. The number of iterations in an activity is governed by complexity, uncertainty, stakeholder feedback, resources, and schedules. In order to satisfy stakeholder expectations, input determines the number of iterations needed, with higher complexity potentially requiring more. Project timetables and resource limitations are considered while balancing efficiency with activity objectives. It focuses on the functionality the system must possess to satisfy the needs of a user (person, external system, device) who interacts with it and manages use cases as the common thread guiding development activities. Through a continuous testing and feedback process, guaranteeing compliance with specific quality standards

Another traditional process is CMMI-DEV (Capability Maturity Model Integration Development), also known as CMMI for Development. Walter Shewhart created the principles of the CMMI model in the 1930s, and in the 1980s, W. Edwards Deming, Phillip Crosby, and Joseph Juran refined them. However, it was in 1989 that Watts Humphrey began using these concepts in IBM and SEI (Software Engineering Institute) (Constantinescu and Iacob, 2007).

CMMI is a software development process improvement model that consists of good practices that address development activities applied to products and services, addressing practices that cover the product life cycle from conception to delivery and maintenance.

The SEI is also in charge of PSP (Personal Software Process), a set of methods created by Watts Humphrey in 1995 for the control and measurement of the estimation of software projects (Pomeroy-Huff et al., 2005). The PSP process helps developers measure and analyze their performance.

In this type of methodology, the importance of system documentation is considered, which allows the software to be understood, extended, and maintained throughout its useful life (Chomal and Saini, 2014). Furthermore, these methodologies provide a well-defined order and structure for software development. However, for these methodologies to work correctly, all development team members require a high degree of discipline, which is even more critical in our Latin culture.

At the beginning of the 1990s, a new type of methodology began to appear: the agile methodology, which presented quick and effective responses to change. It has a flexible project plan and shows simplicity in development.

Agile methodologies are iterative processes; that is, the work is divided into small progressive parts that advance through collaboration with the Client. Through continuous meetings, the process is verified to adjust it appropriately.

These continuous improvement processes promote collaboration between team members, including the Client, who becomes another development team member. Promoting teamwork among clients is part of our effort to create a common goal for the project, enhance communication, and leverage our combined experience to provide excellent outcomes. Including stakeholders in the process encourages proactive opportunity identification, problem-solving, and efficient requirement adaption, improving results and increasing stakeholder satisfaction.

The most prominent agile methodologies in software engineering will be presented below.

Extreme Programming

One of the great representatives of agile practices is XP (eXtreme Programming). Extreme programming is a software development discipline based on the values of simplicity, communication, feedback, courage, and respect (Shrivastava et al., 2021). It was created by Kent Beck in 1996 when he was called to work on the payroll application that Chrysler Corporation was developing.

There, he realizes the shortcomings of traditional methods, the importance of including the Client as another member of the development team, and the benefits of enhancing teamwork.

In extreme programming, each person involved in the project is an important member of the team; even the Client must work with the team during development (Bahrudin et al., 2013).

There are four phases in the XP life cycle:

° Delivery planning phase: In this phase, clients create user stories, and developers take these requirements and convert them into a group of iterations, that is, into a delivery plan (releases).

° Design phase: The XP philosophy of simplicity is emphasized in this design phase.

° Coding phase: In XP, before starting coding, developers must perform unit tests, create the source code, and proceed with refactoring.

° Testing phase: Before coding, unit tests are carried out to constantly test it, and then customer acceptance tests are carried out.

Figure 1. Extreme programming phases

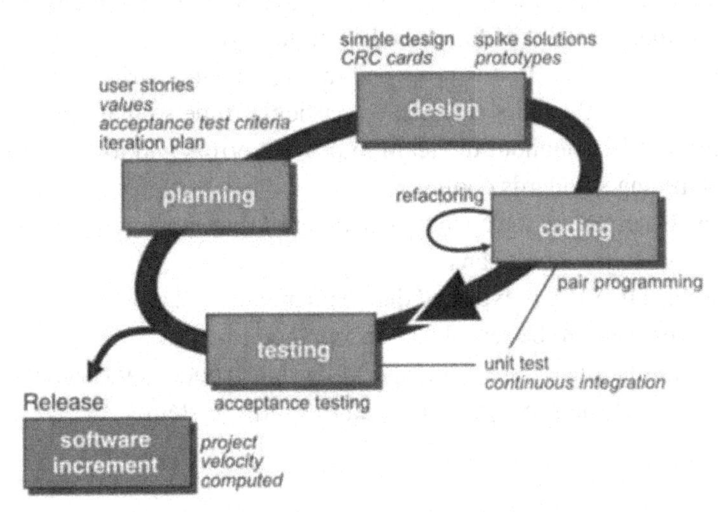

(Purnama, 2023)

The basic principles of extreme programming are as follows:

° Planning game: In this phase, the Client is of great importance since he is in charge of defining the system requirements through documents called user stories, where the Client will express his desires and indicate the system's priorities.

With user stories, the development team determines the system's scope, identifies problems, proposes solutions, and indicates requirements that should be given more importance due to their difficulty or critical point.

° Unit tests: Every time a part of code is implemented, the developer has to write a test. One of the most used methods is test-driven development, known as Test-Driven Development (TDD), where tests are carried out. First, the tests are coded, and then the refactoring of the written code is carried out.

° Minor Releases and Continuous Integration: Continuous integration allows developers to integrate their changes several times daily to create miniature versions. Each version should be as small as possible, contain the most critical business requirements, and make sense.

This process prevents the so-called integration hell from occurring when the project is completed when all developers must integrate the source code and the tedious process of merging and uploading changes to the version control tool. The code must be compiled perfectly when integrating.

° Metaphor: These are brief descriptions of a system's work instead of traditional diagrams and models.

° Simple design: is based on the philosophy that the most significant business value is delivered by the most straightforward program that meets the requirements.

° Refactoring: serves to minimize duplicate code and prepares the system to accept new changes in the future, refactoring is maintaining simplicity, clarity, and the minimum amount of functionality in the code.

° Pair programming: In XP, it is proposed that two people write their code on the same machine; the pairs change at a particular time so that the knowledge is disseminated throughout the team to prevent errors and to extend the use of programming standards coding.

° Collective Ownership of the code: No member of the team owns the code; anyone who wants to modify it can do so.

° 40-hour weeks: In XP, it is expected that team members should not work overtime. Developers need to be rested and encouraged to code.

° Coding standards: Encourage developers to write all code in the same way so that it appears that the same person coded the entire system.

Figure 2. Extreme programming practices

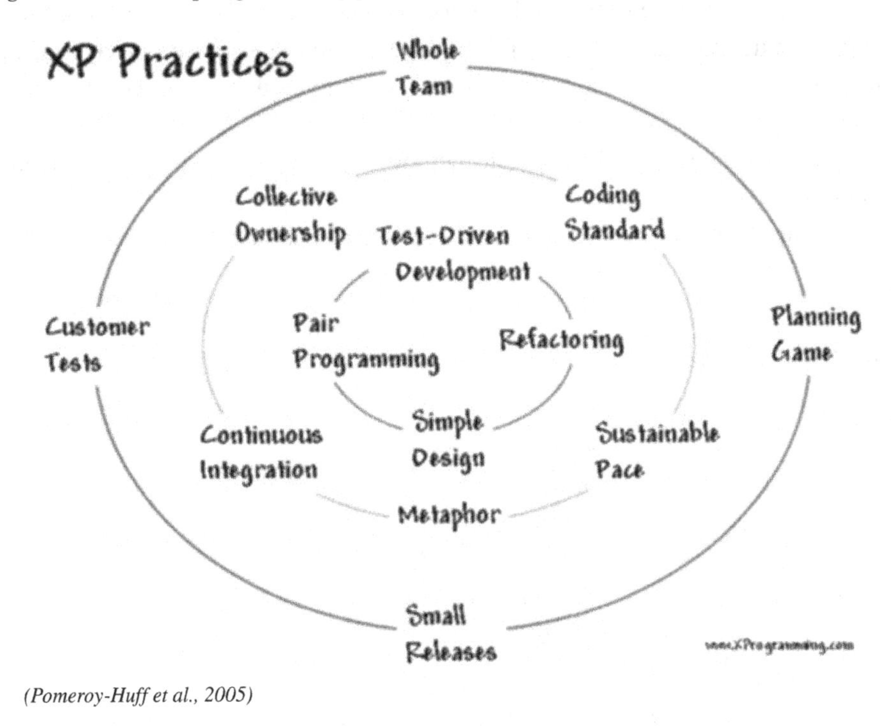

(Pomeroy-Huff et al., 2005)

Scrum

Another exponent of agile practices is Scrum, an agile project development methodology based on a 1986 study by Ikujijo Nonaka and Hirotaka Takeuchi on the best practices for developing technological products of important Japanese companies. However, it was only in 1995 that Ken Schwaber and Jeff Sutherland formally presented Scrum as a software development process (Hema et al., 2020).

Scrum's goal is to improve development feedback to correct errors in the early phases. It is ideal for projects with persistent changes or emerging requirements, such as web projects or products for a new market. Scrum encourages development feedback by giving iterative sprints priority and dividing work into small bits. Daily Scrum sessions monitor development and quickly resolve roadblocks. Stakeholder participation in sprint reviews ensures product alignment and prompt error repair for effective project execution.

As in XP, in Scrum, miniature versions are made in time intervals called Sprints; these cycles are generally one month long. In Scrum projects, teams consider many elements, such as task complexity, market trends, technical issues, risk tolerance, team availability, and organizational restrictions, when determining the duration of sprints. Teams achieve optimal productivity, reduce risk, and guarantee consistent delivery of value by establishing sprint lengths that are balanced. Before each sprint, the work is prioritized, and the tasks that must be delivered at the end of the sprint are selected.

Scrum is a straightforward, adaptable methodology with an incremental structure based on iterations and revisions.

The Scrum team is a self-organized, cross-functional team with three essential roles.

Figure 3. Scrum roles

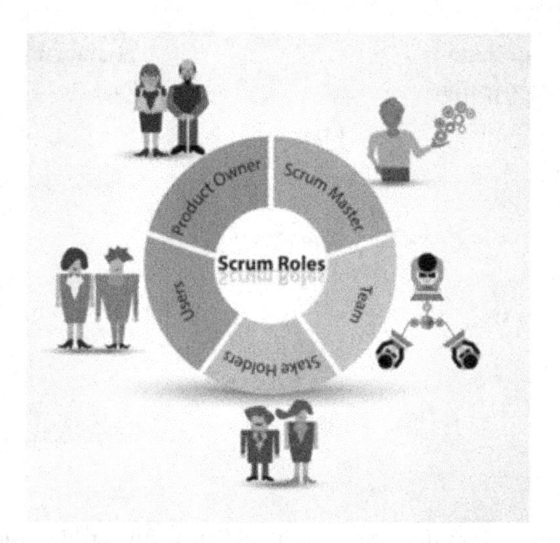

(De Carvalho and Mello, 2011)

- Product Owner: This is the person in charge of creating and managing the product backlog and also ensures that the development team understands the elements in the list.
- Facilitator (Scrum Master): The scrum master ensures that the Scrum process is fulfilled as it should be. Train the development team to be self-organized and multifunctional.
- Development Team: consists of all the people responsible for the delivery of the product; the development team has the power to organize and manage its work; development teams must have the following characteristics:
  - They are self-organized, and no one tells the team how they should do their work.

- They are multifunctional; all team members execute the necessary tasks to deliver the product.
- Scrum does not have titles for development team members - everyone is a developer.
- There are no sub-teams in the development team.

Limiting Agile team size improves communication, collaboration, and decision-making, resulting in faster reactions to change. Smaller teams promote Ownership, responsibility, and efficiency while upholding high-quality standards. This method encourages adaptability, incremental value delivery, and efficient resource utilization.

Components

- Product List (Product Backlog): This list contains all the system requirements; the person in charge of managing it is the product owner, who will ensure that it is constantly updated with the most appropriate characteristics and functionalities.
  The product list is organized by the priority assigned to each requirement, adding greater detail and clarity to the highest-value items.
- Burndown Chart: This graph shows the number of pending items in the product list.
- Sprint Task List (Sprint Backlog): This is a set of product backlog items selected for the sprint.
- Meetings
- Sprint Planning Meeting: This is an eight-hour meeting where the sprint work plan is created, and the activities to be carried out are obtained through the list of products.
- Daily stand-up Scrum: It is a fifteen-minute meeting held every day for each member of the development team to explain the progress of their activities; each person answers the following questions:
  ° What have I done since the last sync meeting?
  ° Was I able to do everything I had planned? What was the problem?
  ° What am I going to do from this moment on?
  ° What impediments do I have, or will I have to fulfill my commitments in this iteration and the project?
- Sprint review: At the end of a sprint, the Sprint review is carried out, where the product owner checks that the delivered products meet the requirements defined in the planning meeting. The development team members also discussed the problems that arose during the sprint. Changes in requirements that cause focus problems and affect the scheduled job delivery are known

as scope creep, and they are a frequent problem during sprints. To resolve these problems and preserve sprint productivity, effective stakeholder management, prioritizing, and communication are essential.

- Sprint retrospective: It is a meeting that is held after the sprint review so that the members of the scrum team can create an improvement plan

During the sprint, visual techniques are used to manage tasks; a board or wall called Scrum Taskboard is used where the status of the requirements is listed, and the objective is to clarify the work. The Burndown Chart is also used, a graph that shows progress and pending work. Scrum uses a Burndown Chart to graphically display work completed and work still to be done. Teams can stay on track and make necessary adjustments with its help since it evaluates plans against actual progress. To complete sprint goals, it encourages accountability, openness, and iterative improvement.

Implementing Scrum increases creativity and makes it easier for the team to respond to feedback. The burndown chart is a critical tool for managing workload, monitoring progress, and identifying problems early on. By visually comparing the progress made against the plan, the tool facilitates prompt modifications, accountability, and motivation. It ensures alignment and a focus on delivering value and promotes cooperative decision-making. It provides a set of simple rules that create a structure for teams to focus on solving challenges and difficulties.

Figure 4. Scrum cycle

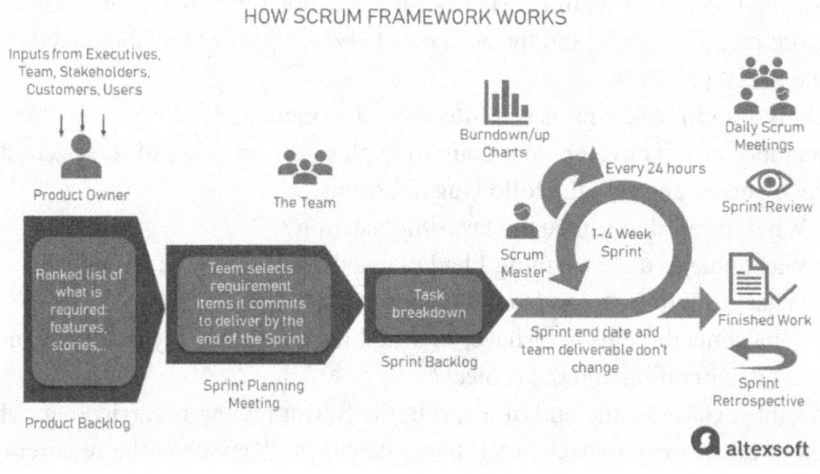

(De Carvalho and Mello, 2011)

Adaptive Software Development

Known in English by its acronym, ASD (Adaptive Software Development), it is a technique created by Jim Highsmith and Sam Bayer in 1998 for the development of complex software. It uses the concepts of adaptive artificial intelligence systems.

This method is iterative and incremental, favoring product development through prototypes. This methodology focuses on human collaboration and team organization.

The main characteristics of ASD are:

- Iterative process.
- Oriented towards the software components rather than the tasks for achieving the objective.
- Tolerant to changes.
- Guided by risks
- The review of components serves to learn from mistakes and restart the development cycle.

This methodology is divided into three phases: speculation, collaboration, and learning.

Figure 5. ASD cycle

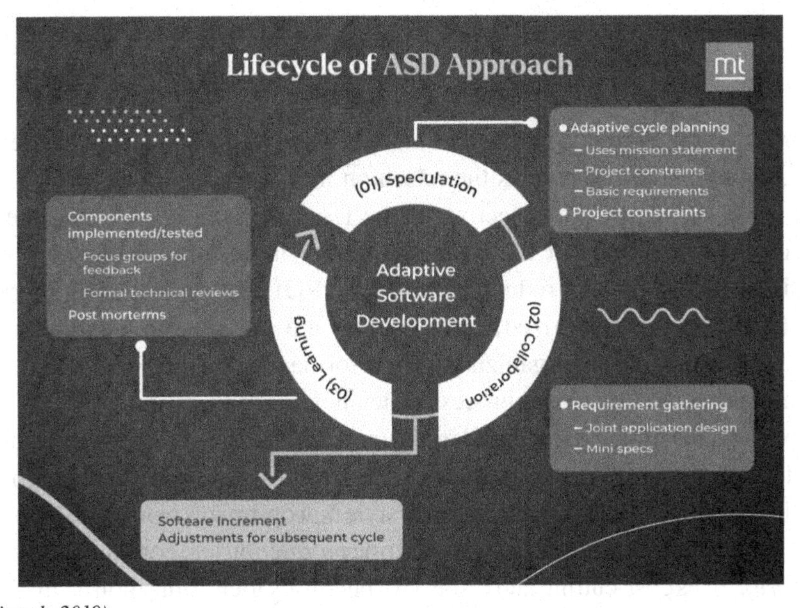

(Qi et al., 2019)

The life of ASD begins with the speculation phase; this is where increments in software development are tentatively planned; in each of these cycles or iterations, you have the option to deviate the development; that is, you can change the order of delivery of the functionalities so as not to delay the project. Teams prioritize feature delivery during the speculating phase by taking into account the value to the Client, business priorities, technical dependencies, feasibility, and risk assessment. Teams optimize the delivery sequence for maximum product impact and success by focusing on high-value features, aligning with business goals, resolving technical dependencies, guaranteeing feasibility, and addressing risks early.

The second is the collaboration phase when the development team members collaborate and communicate with each other to carry out the planned delivery during the speculation phase.

The third phase is the learning phase. In this stage, the quality of the delivery is reviewed, and with this, the aim is to understand better the processes and tools used throughout the development cycle. Four factors are analyzed that will help absorb this knowledge:

- Quality of the result from the Client's perspective.
- Quality of the result from the technical perspective.
- The operation of the development team and the practices it uses.
- The status of the project.

Lean

Before starting a project, companies analyze its viability to know what benefits can be obtained; this is known as the return on investment or by its acronym ROI (Return of Investment); the objective of the Lean methodology is that the ROI analyzed at the beginning of the project is met. With Lean methodology, which emphasizes value and waste reduction, matching ROI with project goals is critical. High-return tasks are prioritized, and resources are allocated to activities tied to outcomes. This improves organizational performance, facilitates well-informed decision-making, and increases process efficiency.

Lean is a management model used in Toyota's production system, whose purpose is to deliver maximum value to customers and avoid waste of resources. Toyota's production method uses a lean management style that optimizes workflow, empowers workers, and reduces waste through just-in-time production and continuous improvement. It prioritizes cost-cutting and respect for people by identifying non-value-added operations, cutting inventory, and improving productivity and quality. It is a work philosophy based on the Japanese cult of Kaizen (continuous improvement), where the only goal is to produce a perfect product that is delivered to the customer.

The concept of Kaizen consists of continuous improvements that, although minor changes, influence the perfection not only of the products but also of the people, "Today better than yesterday, tomorrow better than today."

Although this model began in the textile industry, it was adapted to the automobile industry and is now also adapted to software development. Since it is not a work technique but a way of thinking and acting in an organization, it is also adapted to software development.

Lean's main objective is to focus on a project's cost and ROI and to obtain the quality of the products from their origin; this is reflected in its seven principles (Alaimo and Tortollera, 2013).

→ Eliminate waste
→ Expand learning
→ Decide as late as possible
→ React as quickly as possible
→ Strengthen the team
→ Create integrity
→ See the entire set
→ Eliminate waste: The word waste is described much deeper in Japanese, giving way to three terms:
→ Muri: Avoid overloading processes.
→ Muda: Implies eliminating activities that do not provide value from the customer's perspective.
→ Mura: It means inequality; this implies a lack of regularity in the processes.

Amplify learning: Lean also supports the virtue of learning; through short meetings throughout the development process, the pros and cons of the cycle are provided with feedback along with the Client. These steps help the Client understand his actual needs, and the Developers learn to satisfy these needs.

Decide as late as possible: This premise warns about the need to have all the information available to make decisions and have a better approach with several options when designing to provide a flexible system, avoiding the need to make costly changes after completing the project.

Deliver as quickly as possible: Short development iterations, just like in XP and Scrum, Lean supports fast delivery to get quick feedback and avoid errors.

See the whole: This refers to optimizing the entire system, not concentrating on just the main parts, and remembering the small functionalities, which results in a global vision of the project.

Figure 6. Lean phases

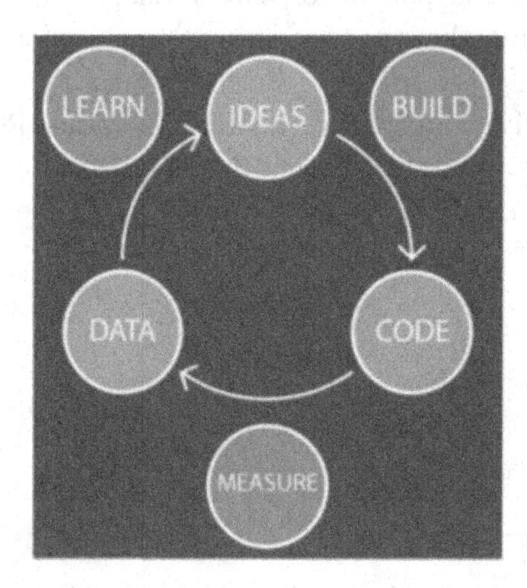

(Dybå et al., 2013)

Agile Manifesto

When studying these methodologies, it can be seen that they have many characteristics in common. This is why, in 2001, the creators of the main agile methodologies decided to unify these aspects, declaring them the principles of the agile manifesto. These are the twelve principles of the Agile manifesto1:

Our highest priority is customer satisfaction through early and continuous delivery of valuable software.
We accept that requirements change, even in the late development stages. Agile processes take advantage of change to provide the customer with a competitive advantage
We frequently deliver functional software, usually within two weeks to two months, with a preference for the shortest possible period.
Business managers and developers work together daily throughout the project.
Projects are developed around motivated individuals. They must be given the environment and support they need and entrusted with executing the work.

- o The most efficient and effective method of communicating information to the development team and among its members is face-to-face conversation.
- o Working software is the primary measure of progress.
- o Agile processes promote sustainable development. Promoters, developers, and users must be able to maintain a constant pace indefinitely.
- o Continuous attention to technical excellence and good design improves Agility.
- o Simplicity, or the art of maximizing the work not done, is essential.
- o The best architectures, requirements, and designs emerge from self-organizing teams.
- o The team regularly reflects on how to be more effective and then adjusts and refines its behavior accordingly.

From these twelve principles, four essential characteristics are extracted that distinguish agile methodologies:

- Individuals and interactions on processes and tools
- Software running on comprehensive documentation
- Client collaboration on contract negotiation
- Respond to change by following a plan

According to Kent Beck, Jeff Sutherland, Jim Highsmith, and the other members who made up the Agile Manifesto, these four characteristics are necessary to define an elegant, lightweight methodology without excessive regulations and values that make the difference from traditional methods. Utilizing theory, practice, and interactive techniques ensures students fully grasp agile concepts. Begin with an overview of agile, highlight important concepts, and provide real-life instances. For real-world preparation, reinforce with practical exercises, promote involvement, and cultivate introspection.

Chapter two proposes a set of practices for software development based on the values outlined in the Agile Manifesto. This proposal serves as a methodological guide that will facilitate the execution of projects developed by members of the research incubators and groups of developers in training and obtaining experience with agile practices in software development. Using Agile techniques significantly improves methodological manuals for developer groups and research incubators throughout training, allowing for ongoing improvement through iterative creation and cooperation. Agile encourages user-centric, successful guidelines by prioritizing high-value features and stakeholder input. This promotes openness and flexibility in continuing improvement. This proposal is guided by a framework based on learning, collaboration, accepting change, the intervention of different actors in the development stages, and the communication of all team members.

METHODOLOGICAL PROPOSAL

Criteria

The extraordinary premise of the Agile manifesto is "Individuals and Interactions over processes and tools." James Shore and Shane Warden explain that the human factor is the primary key in this type of methodology, stating that development Agile is a human art.

People are the most critical resource in a software engineering project; their knowledge and experiences are a significant contribution, and not knowing how to motivate them and discourage their enthusiasm can be detrimental to the project's final result. Foster a supportive culture that values communication and collaboration to keep Agile teams motivated. Recognize accomplishments frequently and allow team members to make decisions, innovate, and improve skills. To maintain involvement and passion, respond quickly to issues, foster work-life balance, and align project objectives.

This is why agile methodologies have caused a stir; the fact that they emphasize the motivation and well-being of people over the perfection of documentation and processes makes skeptics find significant flaws in this type of methodologies.

When implementing an agile methodology in a company, it must be taken into account that each individual has different attitudes and aptitudes towards the same problem; for the aptitude or lack thereof, all agile methodologies handle the concept of retrospective or feedback, a concept that helps development team members improve each time they finish an iteration and can observe their mistakes and retain their qualities. Retrospectives are handled using Agile techniques by holding regular end-of-iteration meetings to examine successes, areas for improvement, and process adjustments. Retrospectives offer a controlled environment for team input, issue resolution, and action item identification. They help Agile teams refine their procedures for better outcomes by encouraging continuous improvement.

To ensure team members have a good attitude toward the change in methodology, it is proposed that these agile techniques be instilled in them from academic training. For now, this guide is aimed at students who are part of research hotbeds since they themselves decided to take part in these types of extracurricular activities.

After the students change their attitude towards the scheme presented by agile methodologies, the teamwork environment must begin to be taught; the members must trust each other and be responsible for ensuring that their tasks are completed on time and without errors. These teams are called self-organized teams.

Agile methodologies seek to ensure that members of self-organized teams have the following values:

- Trust.
- Respect.
- Listen.
- Collaboration.
- Loyalty.
- Discipline.
- Altruism.

Rules for self-organized teams:

- Receive criticism constructively.
- Have a desire to learn.
- Do not look for blame.
- There are no bosses; they are just other development team members.
- Do not look for credit
- Don't make others feel bad.
- Nobody is superior to anyone.
- Do not overshadow others; you must seek the common good.

It is challenging to make all these characteristics work in a group with many members, so agile methodologies suggest that the development team should consist of at most seven people. Agile recommends restricting development teams to seven people to improve communication, collaboration, and production. Smaller teams make coordinating, making decisions, and following Agile principles easier. They stay focused, manage dependencies well, and adjust quickly to changes, resulting in faster value delivery and better outcomes.

According to the study presented by Vikash Lalsing in his article (Lalsing et al., 2012), a team's efficiency often depends on its members' interaction. As the number of members increases, more interaction is required, and it is more complicated to manage them. The more people there are in a team, the more difficult it is to communicate between them. This affects the efficiency and productivity of each one.

Although many agile methodologies based on empirical theory have been created in the last decade, very few companies have been able to implement them adequately. This is first because they involve a change in the company's paradigm and secondly because these methodologies imply profound changes in the strategies and internal behavior of the workgroup.

One of the biggest impediments to implementing this type of methodology is the resistance to change and the bad attitude of the development group members.

Due to this, it is proposed to list a set of agile principles that can be applied in research and software development hotbeds so that students can carry out an experience similar to the one presented in companies when developing a large-scale project and thus improve its performance in an environment of agile techniques.

The objective of combining these agile methodologies is to allow students to obtain the necessary experience with the most important techniques implemented in each of these and to facilitate their understanding until they become a natural part of the processes of the participants of each of them.

The main characteristic of this set of practices is its emphasis on training and preparing students, guiding them to a framework based on learning, collaboration, a good attitude to change, and the intervention of different processes in the development and communication of all team members.

Objectives of the Proposal

In research hotspots and development teams, putting Agile concepts into practice seeks to improve efficiency, creativity, and teamwork. Stakeholder engagement is to be maximized, innovation to be accelerated, and collaboration and openness to be promoted. In research and development projects, agile approaches reduce risks, support hypothesis testing, and encourage long-term progress. The objectives of using this working method in research hotbeds and development groups are the following:

- Encourage students to apply the different methodologies in software engineering. Although the use of these methods is not appreciated in the university environment, they are a great contribution to software development companies that lack this culture.
- Show students the advantages of using a methodology in developing software projects.
- Help seedbed members plan, design, and develop robust systems
- Promote the values instilled by agile methodologies, such as communication with the entire work team, collaboration, trust, respect, and loyalty.
- Prepare participants from an early stage of their academic training for the environments and scenarios they may encounter in the workplace. This strategy

prepares participants for future employment by exposing them to real-world project management methods, cooperation dynamics, and iterative development processes while instilling abilities such as adaptability, effective communication, and incremental value delivery.

Framework

Following the philosophy of agile processes, an iterative and incremental methodology is structured based on collaboration and learning, properties of ADS and Lean, which will also have the same roles as Scrum. This framework consists of two phases that combine techniques from these agile methodologies.

- Planning Phase (Speculation)
- Implementation Phase (Collaboration-Learning)

Figure 7. Methodology phases two

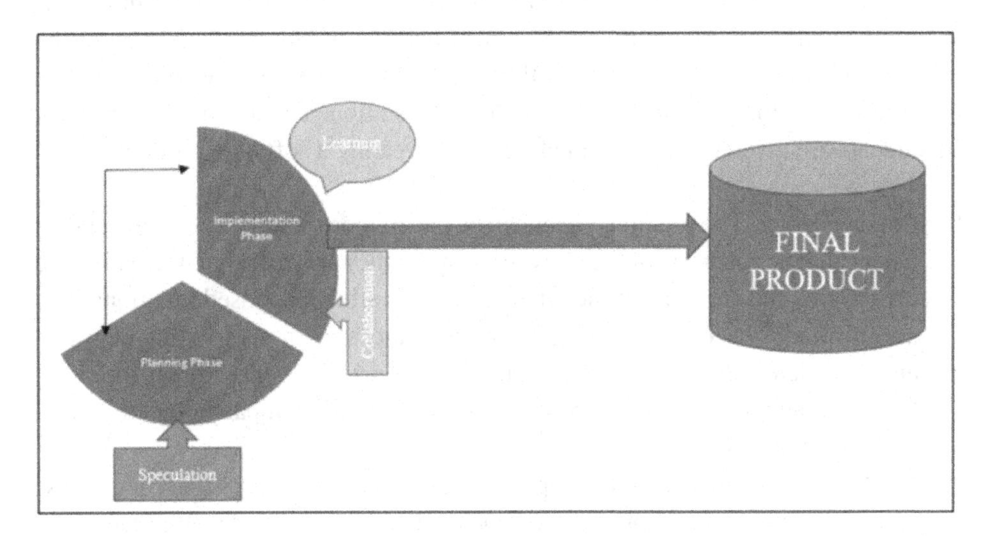

In each of these phases, various activities will be conducted, which in turn will have as many iterations as necessary to achieve the goals defined at the beginning of each activity.

Table 1. Methodology phases three

Phases	Activities	Practices	Artifacts (Deliverables)
Planning	Speculation	User stories	Requirements List (Product backlog)
		Acceptance Tests	
		Make schedule	Schedule
		Solution design	UML diagrams
			GUI prototypes
			Coding Standards
			Frameworks
Implementation	Collaboration and Learning	Coding	Executable test plan
		Unit tests	
		Refactoring	
		Continuous Integration	

Roles

The roles proposed in this proposal are the same as those used in Scrum since, in this methodology, the entire development team is in charge of delivering the software. There are no distinctions or titles between the members; everyone works with the same effort to achieve the objective of iteration. Thanks to this approach, all students are given experience in all areas of creating a software product.

1. Product owner: The product owner can be one or more people who represent the Client and the end users of the software system; they are in charge of creating the user stories and are responsible for ensuring that all the proposed functionalities are fulfilled. The Product Owner's job guarantees alignment with Client and end-user demands by obtaining requirements, prioritizing features, soliciting feedback, and making decisions that correspond with company objectives and user expectations.
2. Facilitator: This person, a development team member, ensures that the process is used as intended and prevents obstacles that prevent the development team from carrying out the activities proposed at the beginning of the iteration. In addition, they will be responsible for transmitting the appropriate knowledge so that students can carry out the assigned activities. Retrospectives are vital in Agile, and they help teams reflect, talk, and learn after each iteration. Team members evaluate performance, identify successes, and collaborate on action plans for improvement, building a culture of continuous improvement, team-

work, and communication. Regular interaction leads to iterative improvements in working habits and product quality.

3. Team: These are all the members of the development team. They will have the same roles in terms of analysis, development, and testing so that they can experience both sides of software development.

Figure 8. Roles four

Phases

Planning Phase

This is the initial phase of the process where the members of the seedbed, with the collaboration of the Client, obtain the scope of the project; in addition, the system requirements are collected through the user stories obtained from the periodic meetings held with the Client. At the outset, role clarification, cooperation, and clear communication are necessary for the seedbed team and the Client to work together effectively. Goals and schedules are synchronized through regular meetings, and collaborative technologies enhance transparency. Setting a solid basis for project success through inclusive communication, attentive listening, and stakeholder input.

When the team members understand the project's scope, they begin by classifying the requirements, prioritizing the essential functionalities, and adding them to the list of functional and non-functional requirements.

With the list of requirements, the delivery planning begins. The requirements with high priority are taken and assigned a delivery date, thus planning the schedule of the first iteration.

Using the Scrum concept of the task board (Scrum Taskboard) and virtual boards, students will organize the activities of each iteration and have better control over each activity.

As the system is defined through speculation, a sketch of the prototype's architecture begins to be designed to analyze the restrictions and risks that may arise throughout the project.

The development tools that best adapt to the Client's needs will also be selected in this phase. Following the technique presented by XP of pair programming, work teams are defined for the exchange of knowledge.

Students who do not know some of these activities or tools will be trained by classmates who already have this type of experience or decide to research the topic and present the knowledge acquired. Of course, this additional teaching-learning time must be added to the project time.

Additionally, coding standards are defined for coding in the database, development tool, and other utilities that intervene in the construction of the system. The aim is that all team members can understand the source code when they need to modify functionalities that have not been made by them.

List of products of the first phase:

- List of functional and non-functional requirements: This document is maintained by the facilitator and obtained through User Stories. It is like a contract between the Client and the Development Team regarding what is going to be built. Clear communication is essential to obtaining system requirements, user stories, project goals, and user demands. While precise technical explanations control expectations, active listening facilitates comprehending the Client's vision. Building trust and providing client-centric solutions through cooperation and feedback ensures project success.
- Meeting and version delivery schedule: The facilitator will also manage the schedules, and based on the stipulated dates, the delivery time of the versions will be managed.
- Development tools: The tools that best adapt to the proposed architecture will be chosen depending on the architecture selected for the project.
- Coding standards: Based on the development tools, team members will establish coding standards used throughout the coding process.

Implementation Phase

In the second phase of the process, the development team defines the software architecture, considering the requirements taken in the first phase. The XP feature is evoked to retain simplicity in design by avoiding duplication in logic and minimizing the number of methods to satisfy customer needs. In Extreme Programming, avoiding logic duplication keeps design simplicity, lowers complexity, and improves clarity. By removing unnecessary code, programmers can create software that is easier to maintain and modify, making debugging more accessible and improving the codebase over time for higher-quality systems.

The system must be easy for the user to use, flexible, highly scalable, and adaptable to changes at any stage of the software's life.

Having a complete image of the system, we begin with designing the user interfaces (GUI), looking for a strategy that satisfies the Client's specific needs and facilitates its maintenance over time.

The design of the database structural model is also carried out to support the system and business logic that has been proposed up to this stage.

In this phase, coding begins using the coding standards defined in the planning phase. In this phase, the concept of refactoring comes in, another of the essential characteristics of agile methodologies. Research articles must highlight the benefits of restructuring to reduce unnecessary code. Refactoring reduces errors and duplication, improves manageability, and increases developer productivity for debugging and future improvements. It also encourages standardization and consistency in the code, which results in better software and lower technical debt.

After implementing a new feature to the system, we look for a way to make it as simple as possible without losing functionality. This process is called refactoring. This, although it involves a more significant effort and investment in the development stage, allows us to minimize duplicate code, adopt best practices, implement new technologies, and prepare the system to receive the changes necessary over time in the most transparent way possible of life.

In this phase, as an additional task to the development process, the first test application is carried out. The developers execute these applications to test and certify the modifications and new functionalities injected in the current development stage.

Once the modifications and newly implemented functionalities are certified, the development team integrates the improvements and changes made to implement the functionalities, delivering each development section for subsequent review and certification.

Finally, based on the deliverable made during the integration, the final tests of the added functionalities are carried out, with which it is sought that the modifications generated meet and satisfy all the requirements that were intended to be resolved during this stage.

Additionally, it is possible that during the execution of these stages, new requirements or changes in the needs initially expressed by the Client are identified, for which the process begins again from the planning stage.

Second Phase Product List

- Test plan: Test cases contain the specifications for validating the system. They are carried out based on the use cases and propose all possible scenarios that they may contain.
- Executables: This artifact includes the source code executable and the deployment scripts necessary for installing the product in the customer environment.

When including source code executables and deployment scripts, it is critical to ensure consistency and stability across various environments. Thorough testing of deployment scripts in numerous contexts eliminates mistakes during installation, ensuring a smooth deployment procedure for clients.

This chapter describes the process from the research hotbeds for developing the first Agenda module, UNotes, of the "UPB Mobile" project. This process is the first pilot of the agile practices proposed in this work and was carried out by the Urban Communication Research Groups assigned to the Faculty of Social Communication and the Research, Development, and Application Group in Telecommunications and Computing group assigned to the Faculty of ICT Engineering. The effectiveness of piloted Agile practices in the project was evaluated using sprint reviews, retrospectives, stakeholder surveys, velocity tracking, and analysis of KPIs. These methods yielded insights into Agile methodologies' effectiveness in achieving project objectives, delivering value, and fostering continuous improvement.

APPLICATION IN RESEARCH SEEDHEAD

This project sought to identify the agile practices that best adapt to the training of students from research incubators and software development groups. Due to a lack of exposure and training, reliance on traditional project management, resource limits, local success stories, cultural hurdles, and a lack of Agile mentors, incubator members encounter difficulties using Agile software development approaches. All of these obstacles make it more difficult for incubators to implement Agile. Investigate the techniques of the most representative agile methodologies of software engineering, which will include practices for managing various expectations and the high percentage of uncertainty associated with formative research projects.

These criteria were decisive in investigating defined and documented agile processes such as extreme programming XP, SCRUM, ASD, and Lean. This allowed us to find important guidelines to foster the culture of collaborative work, promote autonomous learning, and apply strategies for interdisciplinary work.

The UPB Mobile project began as a formative investigation of social communication students' use of mobile devices at the University. As a result of this process, the opportunity was identified to take advantage of this survey of scenarios linked to mobile applications and thus try to capitalize on the needs and expectations that were expressed by students and teachers in prototype development.

The characteristics associated with the UPB Mobile project, long-term development, increasing complexity, and the possibility of enhancing in a single application the different systems that exist at the University, the Digital Communication Incubator and the Internet of Things Incubator take on the challenge of leaving a contribution to the University with the development of the software prototypes necessary to carry out the project, and in this sense, it seeks to implement a strategy that allows formalizing the documentation and the development process. While the IoT Incubator helps startups create IoT devices, sensors, and solutions, the Digital Communication Incubator promotes innovation in social media platforms, messaging apps, and digital marketing tools. At the same time, they serve different digital technology domains and businesses, and both foster technical innovation.

The Members of the Internet of Things hotbed decide to track and apply the agile practices studied in their Software Engineering courses of the Systems Engineering and Computer Science program of the Faculty of Engineering for the development of a prototype and start under this methodological framework for the development of the UPB Mobile project. Teamwork, time management, problem-solving, flexibility, and practical experience with industry tools are all enhanced when Agile practices are incorporated into student education. This approach prepares students for employment in project management and software development by fostering incremental improvement and mimicking real-world settings.

The components of the strategy used to develop the UNotes application within the framework of the UPB Mobile project are described below.

Roles

Product owner: The product owner was represented by students from the Digital Communication incubator assigned to the GICU group of the UPB Faculty of Social Communication and Journalism, whose members were in charge of defining, specifying, and validating the project requirements.

Facilitator: The person in charge of providing the students with the necessary tools for developing the project and establishing the team's work dynamics, as well as the dialogue between seedbeds in this project, was Engineer Oscar Eduardo Sánchez. The facilitator is crucial to the project's success since they lead the team, promote collaboration, and guarantee that Agile methods are applied correctly. They support efficient workflow and optimize the advantages of the Agile process by helping to overcome challenges, settle disputes, and align with stakeholders.

Team: From the second semester onwards, the development team was made up of Systems Engineering and Computer Science students with varying degrees of experience in the project technologies. They were members of the Internet of Things hotbed, performing analysis, development, and testing tasks.

Planning Phase

The students of the Internet of Things incubator began periodic meetings with the members of the Digital Communication incubator to recognize the scope of the UPB Mobile project through requirements survey, specification, and estimation techniques supported by user stories. As they carried out the task of defining the requirement list with the product owner, they held training workshops on application development tools for mobile devices. A mobile app development tools workshop can be effectively organized by determining the target audience and their degree of expertise, choosing pertinent tools, and creating a well-structured curriculum. Facilitating participant input and resource access aids in assessing efficacy and improving subsequent sessions.

Some of the members of the IoT Seedbed, with a greater degree of expertise in using tools, presented the development environments and licensing forms to the other members of the seedbed to identify which one best suited the project. Workshops were held with the first free tool for developing mobile device applications using Web technologies called Phonegap. This framework allows the construction of cross-platform apps; that is, the prototype to be created can be used on operating systems such as iOS, Android, BlackBerry OS, Windows Phone 7, or Symbian. Agile approaches for prototyping attempt to quickly evaluate concepts, obtain feedback, and reduce risks at an early stage. Before full-scale development, prototypes allow for iterative improvement and facilitate stakeholder interaction while guaranteeing alignment with business objectives and user needs. The students also researched Kanban boards for managing team activities. These tools are task control tables widely used in methodologies such as Lean and Scrum. The tool chosen for its usability and licensing was Trello, a web application that allows you to store tasks on a virtual board. Strategies for ensuring usability and adaptability in GUI design include user research, user-centered design principles, straightforward navigation, clear feedback/error messages, responsive design, and user customization possibilities. This allowed the members of the hotbed to control the tasks assigned in the face-to-face meetings using free and virtual access.

After specifying the requirements with the social communication seedbed, the Internet of Things, members classified the highest priority functionalities and designed the user interfaces approved by social communication.

After speculating on the development environments and techniques for managing user stories, the requirements were specified with the Social Communication Seedbed. The requirements specification document (Product Backlog) was the input for members of the Internet of Things hotbed to classify functionalities, prioritize tasks, and estimate effort. This was accompanied by a requirements validation strategy called prototyping graphical user interfaces.

Implementation Phase

The UPB Mobile project will be considered the institutional app; the graphic proposal of the system must have colors, typologies, and images aligned with the University's corporate image. This forced the team members to seek support from the UPB communications department for the graphic design of the application. The project team must make a formal request to the UPB communications department together with the scope, needs, and deadline information to receive graphic design support. The department works with the team to ensure feedback and communication while evaluating feasibility and designating a designer. After that, final assets are sent for the application to integrate with.

Regarding the design of the system, data modeling and the definition of process and block diagrams were carried out to support the definition of the architecture.

The students also proceeded to define the architecture of the first module Client, Server architecture, with three layers:

- Front End: This layer is in charge of specifying the system components that will be executed on the client side (Mobile Device), taking as a starting point the processing, storage, screen sizes, and loading restrictions of these devices.
- Services Layer: This layer is in charge of data management and transport from the Client to the server and vice versa. It includes a set of functionalities exposed in the cloud, such as authentication, registration, and notifications.
- Data Access Layer: This layer stores information that circulates through the system. It includes a database management system (DBMS) that supports the relational model for information organization.

Data normalization for integrity, indexing for retrieval efficiency, transaction management for consistency, query optimization for performance, security protocols for privacy, and support for relational algebra operations like join, select, and project are some DBMS supporting factors for the relational model. Finally, the team proceeded to the implementation phase to carry out coding, testing, and integration in 30-day development cycles, which allowed the construction of incremental versions of the product that, in turn, were validated by the students of the Digital Communication Hotbed in each delivery.

The prototype developed was piloting a set of agile practices within the framework of the proposal presented in this work. The project's evaluation of the piloted Agile methods comprised tracking delivery schedules, deliverable quality, stakeholder satisfaction, teamwork, flexibility, and process enhancement. These observations assess the efficacy of Agile approaches in accomplishing project objectives, providing value, and promoting a collaborative and continuous improvement culture. This piloting was proposed to find and affirm the techniques worked on and test them for subsequent reviews and adjustments of the methodological guide applicable to research incubators and groups of developers in training.

CONCLUSION

There are various agile methodologies that allow managing, administering, and planning software developments. However, the characteristics of the members in some work teams and their operating conditions require adapting processes that allow articulating practices from various sources of information. Hybrid techniques can be adapted to adjust to the different needs presented by software projects.

This exercise sought to generate alternative mechanisms for selecting methodological strategies by characterizing the group of developers. A flexible methodological design is proposed without ignoring the rigor associated with the choice of the methodology to be adopted in a project, and finally, a proposal for the practical application of a set of activities to be implemented in search of enhancing the characteristics of each one and articulating them for the achievement of better results.

Also, we wanted to provide the student members of the research groups with a set of techniques tested in different software development methodologies, which can be applied in formative research projects and work challenges that they may encounter in their performance.

Based on the experiences lived in this research and validated in professional practice, it is concluded that the importance given to the type of methodology applied to a software project is transferred to the attitude and disposition of the development team members. Human factors are the most influential variables for a successful outcome.

Finally, emphasis is placed on the importance of continuous learning, collaboration, and communication, as well as the skills and competencies that development team members must acquire to carry out a software project.

REFERENCES

Al-Ratrout, S., Tarawneh, O. H., Altarawneh, M. H., & Altarawneh, M. Y. (2019). Mobile application development methodologies adopted in Omani Market: A comparative study. *International Journal of Software Engineering and Its Applications*, 10(2).

Ayada, W. M., & Hammad, M. A. E. E. (2023). Design quality criteria for smartphone applications interface and its impact on user experience and usability. *International Design Journal*, 13(4), 339–354. 10.21608/idj.2023.305364

Bahrudin, I. A., Mohd Hanifa, R., Abdullah, M. E., & Kamarudin, M. F. (2013). Adapting extreme programming approach in developing electronic document online system (eDoc). *Applied Mechanics and Materials*, 321, 2938–2941. 10.4028/www.scientific.net/AMM.321-324.2938

Chomal, V. S., & Saini, J. R. (2014). Significance of software documentation in software development process. *International Journal of Engineering Innovations and Research*, 3(4), 410.

Constantinescu, R., & Iacob, I. M. (2007). Capability maturity model integration. *Journal of Applied Quantitative Methods*, 2(1), 31–37.

De Carvalho, B. V., & Mello, C. H. P. (2011). Scrum agile product development method-literature review, analysis and classification. *Product: Management & Development*, 9(1), 39–49. 10.4322/pmd.2011.005

Dybå, T., Dingsøyr, T., & Moe, N. B. (2014). Agile project management. *Software project management in a changing world*, 277-300.

Hema, V., Thota, S., Kumar, S. N., Padmaja, C., Krishna, C. B. R., & Mahender, K. (2020, December). Scrum: An effective software development agile tool. *IOP Conference Series. Materials Science and Engineering*, 981(2), 022060. 10.1088/1757-899X/981/2/022060

Lalsing, V., Kishnah, S., & Pudaruth, S. (2012). People factors in agile software development and project management. *International Journal of Software Engineering and Its Applications*, 3(1), 117–137. 10.5121/ijsea.2012.3109

Pomeroy-Huff, M., Mullaney, J. L., Cannon, R., & Seburn, M. (2005). *The Personal Software Process (PSP) Body of Knowledge, Version 1.0.*

Purnama, I. (2023). Clinical Information System Using Extreme Programming Method. International Journal of Science. *Technology & Management*, 4(5), 1229–1235.

Qi, E. S., Bi, H. T., & Bi, X. (2019). Study on agile software development based on scrum method. In *Proceeding of the 24th International Conference on Industrial Engineering and Engineering Management 2018* (pp. 430-438). Springer Singapore. 10.1007/978-981-13-3402-3_46

Shafiee, S., Wautelet, Y., Hvam, L., Sandrin, E., & Forza, C. (2020). Scrum versus Rational Unified Process in facing the main challenges of product configuration systems development. *Journal of Systems and Software*, 170, 110732. 10.1016/j.jss.2020.110732

Shrivastava, A., Jaggi, I., Katoch, N., Gupta, D., & Gupta, S. (2021, July). A systematic review on extreme programming. *Journal of Physics: Conference Series*, 1969(1), 012046. 10.1088/1742-6596/1969/1/012046

Chapter 2
Building Blocks:
Meshing

Harikumar Nagarajan
Global Data Mart Inc., USA

B. Tamilarasan
Madurai Kamaraj University, India

M. Thanjaivadivel
REVA University, India

ABSTRACT

This work specifically examines the analysis of building blocks structures that experience finite displacements. The analysis was conducted with incremental sequential techniques. The Newton-Raphson iteration approach is used to load structural stages that have the ability to either soften or harden. A modified Newton-Raphson iteration technique is employed for loading stages in which the determinant of the global stiffness matrix is near 0 or negative, as observed in the snap-through scenario. The advanced computer language Fortran is used for structural analysis. To ensure the program's integrity, SAP (structural analysis program) is used for verification. The software is then used in several structural systems case studies. To accomplish this, the authors compare the axial stiffness of panels and jack elements as well as the height-to-span ratios of various structures. The structural responses were modelled using stiffening, snap-through, and softening load-displacement curves.

DOI: 10.4018/979-8-3693-3964-0.ch002

INTRODUCTION

Linear elastic systems have several facilitating properties. In linear elastic systems, the proportional relationship between loads and responses ensures predictability and simplicity, enabling easy determination of varying loads using introductory algebra and calculus. This linearity allows for efficient design optimization, as changes in load directly correlate with structural behavior. It also simplifies validation and verification, enhancing safety and reducing unexpected failures. First, there is a proportional relationship between external influences (loads) and structural responses (reaction forces or displacements) or between the responses of reaction forces and displacements. Second, for linear elastic systems, the law of superposition applies, where the total response of the structure is the algebraic sum of the structure's reactions to the combination of loads. In addition, based on linearity and the application of the law of superposition, the sequence of load combinations in a linear elastic system does not affect the final response of the total structural system. It is said that a linear elastic structural system is a structural system with a short memory. (Jagota Vishal, et al, 2013)

In the case of a nonlinear system, both force-displacement proportionality properties, the law of superposition does not apply, and the loading sequence dramatically influences the final result of the structural response. The lack of force-displacement proportionality complicates prediction and analysis in nonlinear structural systems, requiring advanced numerical methods. Nonlinear behaviors like stress redistribution, path-dependent responses, and bifurcations challenge design and safety assessments. This demands sophisticated tools and comprehensive testing to ensure reliability and performance. Thus, technical solutions that can be applied to linear elastic systems (one-time solutions) cannot directly apply to nonlinear structures. A nonlinear elastic structural system is a system with a long memory.

Pepper Darrell et al. (2005) describe the sources of nonlinearity in structural systems as consisting of two types: material nonlinearity and geometric nonlinearity, although both can occur simultaneously. Material nonlinearity arises from a stress-strain relationship that is not linear, both inside and beyond the elastic range (elastoplastic or perfectly plastic). Geometric nonlinearity arises from the arrangement and interconnection of the structural system, as well as tiny displacements. However, the main focus of this work will be solely on the aspect of infinitesimal displacement.

Nonlinear problems are usually solved using sequential incremental solutions, namely load-displacement relationships that change for each increase in load. Sequential incremental solutions for nonlinear structural problems break complex responses into manageable steps, enhancing control and accuracy in capturing material behavior and stress redistribution. This method identifies critical points like yielding or buckling and accommodates various material models and loading conditions. It is ideal for path-dependent problems and complex geometries that are challenging to direct methods. This method algorithm uses a secant or tangent modulus approach in the computing process. The secant modulus approach uses a straight line from the origin to the current stress-strain point, offering simplicity and stability for large deformations but less local accuracy. Using the current slope, the tangent modulus approach captures detailed nonlinear behavior more precisely but requires more computational effort and frequent updates.

The scope of discussion of this research is limited to elastic structures for building block cases with finite displacements. Incremental sequential techniques enhance the analysis of structures with finite displacements by breaking down nonlinear behavior into manageable steps and accurately tracking geometry, material properties, and load distribution. This iterative approach updates the configuration and stiffness matrix at each step, improving convergence and stability. It allows precise assessment of responses, identification of critical points, and practical design optimizations. The solution method used is an incremental approach. The analysis was carried out using modified Newton-Raphson iterations and jacking techniques. The load increase was set to be relatively small and cumulative until a balanced configuration was obtained based on the load intensity under consideration. The structural modeling to be analyzed is applied to joint frame structures.

The aim of this research is to find a method to develop a method and numerical model that can be applied in the analysis of elastic nonlinear structures. In addition, structural problems with infinite displacement conditions that cannot be solved using exact methods can be approached using numerical methods.

METHOD

Finite Element Method

According to (Haghani Chegeni et al., 2022), the finite element formulation is based on elasticity theory and the working principle of virtual displacement. Understanding elasticity theory aids in designing and optimizing structural systems by analyzing material deformation under various loads. It predicts stress distribution, strain, and displacement, ensuring structural integrity and material efficiency. This knowledge supports advanced methods like finite element analysis, enabling precise simulations and iterative design improvements for durability and resilience. The theory of elasticity is the basis for formulating a structure's force and displacement mechanisms. The structures under discussion are limited to bodies or systems that are continuous and made from elastic, isotropic, and homogeneous materials. The discussion begins with an overview of the basic concepts of theory regarding stress, displacement and strain relationships, stress and strain relationships.

From the stress theory, we discuss the force balance equation for a small element. Still, this balance equation is unsuitable for preparing balance equations in the formulation of the finite element method. The force balance equation isn't suitable for the finite element method as it needs to account for individual element deformations. While the finite element method discretizes structures into behavior-governed elements, the force balance equation addresses overall structural force equilibrium, ignoring internal forces and element deformations. Using the force balance equation directly in the finite element method would inaccurately represent individual element behaviors and interactions. Another method discussed here is the balance equation based on the principles of energy or work.

The principle of virtual displacement approaches the definition of balance. This method is becoming more common because of its superiority over the minimum potential energy principle, namely that the virtual displacement principle does not depend on whether the stress-strain relationship is linear and applies to non-conservative structures (systems that do not have a potential function). Conservative structures maintain energy conservation by balancing potential and kinetic energy during motion, while non-conservative structures deviate due to dissipative forces like friction or damping. This leads to energy changes over time, contrasting with the constant mechanical energy of conservative structures. The virtual displacement principle surpasses the minimum potential energy principle in finite element formulation by providing greater flexibility for complex boundary conditions and non-conservative forces. It satisfies equilibrium and compatibility conditions, making it ideal for dynamic loading and nonlinear material behavior while allowing direct incorporation of external forces and moments. This makes it more versatile and robust for detailed modeling of complex interactions and boundary conditions. The principle of virtual displacement states that for equilibrium, the work done by internal forces during any virtual displacement equals the work by external forces. This principle ensures compatibility and equilibrium, aids in formulating stiffness matrices and force vectors in finite element analysis, and helps assess stability and failure modes for accurate and reliable designs.

A finite element formulation can be prepared after obtaining the balance criteria from the virtual displacement principle.

Axial Bar Finite Element Formulation

According to (Li Kun et al., 2017), components are straight rods that experience axial variations due to external loads and end conditions such that only various axial deformations occur in the rod. So, for example, the external force acting is axial to the rod, and the two ends are such that they do not transmit the moment.

Pendel is the simplest type of finite element. Therefore, we will also start learning finite elements by discussing these elements before we are ready to discuss more complicated and complex types of finite elements.(Langlois William, et al, 2021), (Hui Long, et al, 2021). Finite element displacements directly impact structural analysis by influencing the calculation of strains, stresses, and internal forces. Accurate displacement predictions determine deformation under loads, affecting force distribution and stability assessment. Precise modeling and computation of element displacements are crucial for reliable and safe structural analysis.

a) stem and style
b) tip force and displacement

(c) internal force

Figure 1. Axial bar force and displacement

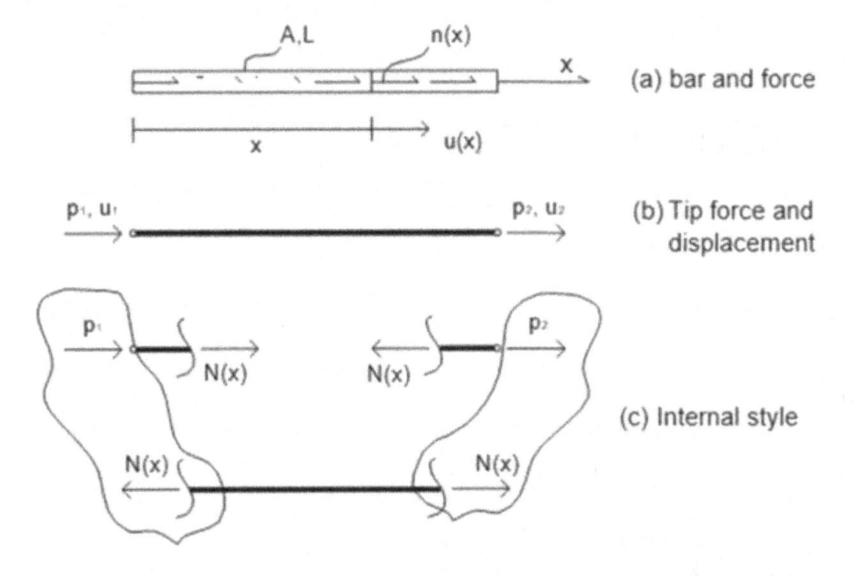

First, we consider a straight bar bounded at both ends by 1 (x_1) and $2(x_2)$ so that the length is $L = x_1 - x_2$ and the cross-sectional area is A. The displacement of the rod is in the axial direction, and we call it $u(x)$, which will be taken as an interpolation of the tip displacements arranged in the element displacement vector

$$\{\hat{u}\} = \{u_1, u_2\} \tag{1}$$

which is related or corresponds to the element end force vector

$$\{\hat{p}\} = \{p_1, p_2\} \tag{2}$$

Displacements are interpolated by writing

$$u\left(x\right) = \sum_{i=1}^{1=2} N_i(x)\, u_i = [N_1(x)\ N_2(x)]\{u_1, u_2\} \tag{3}$$

Since the node displacement has two components, we use the assumptive displacement in terms of the linear polynomial.

$$u(x) - [1, x]\{\alpha_1, \alpha_2\} = [N_a]\{\hat{\alpha}\} \tag{4}$$

The vector $\{\hat{\alpha}\}$can be expressed as a vector$\{\hat{u}\}$ by considering the displacement boundary conditions

$$u(0) = u_1 \ ;\ u(L) = u_2 \tag{5}$$

which, if plugged into Equation (3) above, gives

$$\begin{bmatrix} 1 & 0 \\ 1 & L \end{bmatrix} \begin{Bmatrix} \alpha_1 \\ \alpha_2 \end{Bmatrix} = \begin{Bmatrix} u_1 \\ u_2 \end{Bmatrix} \tag{6}$$

Or.

$$[A]\{\hat{\alpha}\} = \{\hat{u}\} \tag{7}$$

the inversion of the above Equation gives

$$\begin{Bmatrix} \alpha_1 \\ \alpha_2 \end{Bmatrix} = \begin{bmatrix} 1 & 0 \\ -1/L & 1 \end{bmatrix} \begin{Bmatrix} u_1 \\ u_2 \end{Bmatrix} = [A]^{-1}\{\hat{\alpha}\} \tag{8}$$

which, when plugged into Equation (and by comparing the result with the form in Equation (3), yields a function of the form

$$\left[N_1(x)\ N_2(x)\right] = \left[\left(1 - \tfrac{x}{L}\right)\ \left(\tfrac{x}{L}\right)\right] \tag{9}$$

With the displacement function arranged, the process of constructing the stiffness matrix begins by determining the strain vector given by

Figure 9 illustrates how structural enlargement relates to stiffness reduction, highlighting nonlinear structural softening. Increased deformation or loading induces localized yielding, decreasing overall stiffness, which is particularly noticeable in metals or reinforced concrete. Understanding this relationship is vital for predicting structural response accurately and ensuring system safety and reliability under varied loading conditions.

$$\left\{\hat{\varepsilon}\right\} = \left\{\varepsilon_{xx}\right\} = \frac{\mathrm{d}}{\mathrm{dx}}\left[N_1(x)\ N_2(x)\right]$$

$$\left\{\varepsilon_{xx}\right\} = \left[B\right]\left\{\hat{u}\right\} = \left[\tfrac{-1}{L}\ \tfrac{+1}{L}\right]\left\{\hat{u}\right\} \tag{10}$$

and construct a stress vector

$$\left\{\hat{\sigma}\right\} = \left\{\sigma_{xx}\right\} = \left[E\right]\left\{\hat{\varepsilon}\right\} = \left[E\right]\ \left[B\right]\left\{\hat{u}\right\} \text{ (Eq.11)}$$

with the composition of the material stiffness matrix [E] and matrix [B], the pendel element stiffness matrix can be prepared by applying Equation (12)

$$\left[k\right] = \iiint_V [B]^T [C]\ [B]\ dV \text{ 12}$$

produce,

$$\left[k\right] = \int_0^L [B]^T \left[E\right]\ \left[B\right] AdX = \begin{bmatrix} EA/L & -EA/L \\ -EA/L & EA/L \end{bmatrix} \text{ 13}$$

It has been explained previously that in the case of pendel elements, external forces are only axial to the element. If it is stated that the external force is axial in $n(x)$, then the external force vector can be represented by interpolating the force intensity at the vertex

$$\left\{\hat{n}\right\} = \left\{n_1 n_2\right\} \text{ (Eq.14)}$$

so that the axial external force $n(x)$ is approximated by using the shape function as an interpolator so that

$$n\left(x\right) = \sum_{i=1}^{1=2}[N_1(x)\ N_2(x)]\left\{n_1 n_2\right\} = \left[N\right]\hat{n}\Big\} \tag{15}$$

Thus, the equivalent end force vector of the element can be constructed

$$\left\{\ddot{n}\right\} = \int_0^L [B]^T\left\{\hat{n}\right\} dx = \begin{bmatrix}(2n_1 L/3 + & n_2 L/3)\\ (n_1 L/3 + & 2n_2 L/3)\end{bmatrix} \tag{16}$$

Nonlinear Analysis

Complex structural systems can exhibit nonlinear behavior in geometry and materials under static and dynamic loading. Structural nonlinearities stem from geometry and materials interaction in static and dynamic loading. Large displacements cause geometric nonlinearities, while material nonlinearities come from stress-strain behaviors like yielding. Dynamic loading introduces further complexity with time-varying loads, resulting in intricate structural responses such as buckling and chaotic vibrations. Material nonlinearity is characterized by structural elements that deform beyond their elastic capacity. Meanwhile, geometric non-linearity can occur even if the material behavior is still in the linear elastic region and the load is below average due to an increase in the tiny displacement of the structure due to a rise in loading (large displacement). Geometric nonlinearities can notably affect Y-direction nodal points, impacting around 13.2% compared to SAP output. These arise from large displacements, rotations, and deformations, altering stiffness and load distribution, leading to behaviors like buckling or large deflections. Considering these nonlinearities is crucial for precise structural analysis, ensuring system reliability and safety under diverse loading conditions. (Karan Kumar Pradhan, et al, 2019), (Haghani Chegeni et al., 2022), (Kythe Pk, 2004) Geometric nonlinearity, even with linear elastic material behavior, impacts structural analysis by altering stiffness matrices, stress distributions, and stability. It causes localized stress concentrations, especially in slender or curved elements, not captured in linear analysis, potentially leading to buckling. Considering geometric nonlinearity is vital for accurate structural behavior predictions.

In nonlinear analysis,(Nagarajan Praveen et al., 2018) emphasized that the structure will show softening or stiffening behavior when the initial loading is carried out. However, in between, the structural behavior can experience a snap-through before transitioning to stiffening behavior. More details can be seen in Figure 2. Geometric imperfections, material properties, loading conditions, and boundary constraints influence snap-through behavior in structural systems. Geometric imperfections, such as initial deflections, alter stress distribution, while material properties dictate the structure's response. Careful consideration of loading magnitude and boundary constraints is crucial to mitigating snap-through risks in designs.

To handle such conditions, the modified Newton-Raphson method will be used where this method has proven to be quite effective for handling nonlinear problems and has been widely used by previous researchers(Fei Yi-Fan et al., 2022), (Li Gang et al., 2018), (Pierfrancesco Cacciola et al., 2021).

The arch length method or jack method can be utilized for softening and stiffening situations and for snap-through conditions. The arch length method in structural analysis traces the equilibrium path in non-linear issues like buckling. It incrementally adjusts loads while keeping a constant arc length in the load-displacement space, capturing behaviors such as snap-through and snap-back. This allows continued analysis beyond critical points to understand structural performance under varying loads comprehensively. The Jack Method reveals specific structural behaviors and local weaknesses that commercial software may overlook by incrementally applying loads and observing failures. It complements commercial software, leading to more accurate analyses, optimized designs, and improved safety. This combined approach enhances the understanding of structural performance under complex loading conditions.

Figure 2. Representative behavior of nonlinear structures

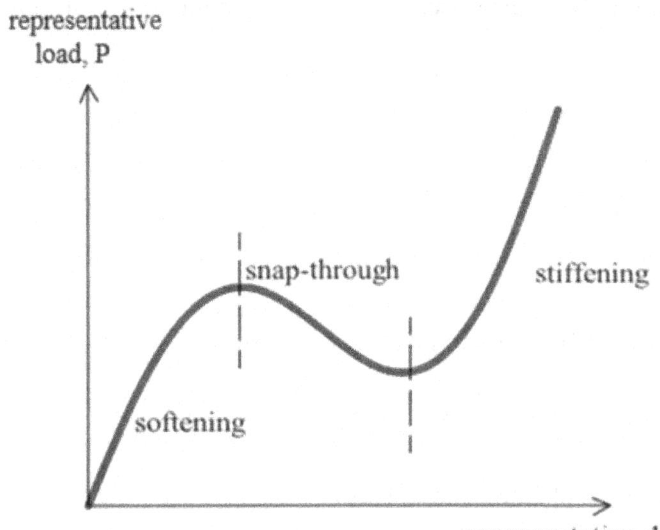

The solution technique uses an incremental method, where the displacement at time t gets incremental over time additional Δt, so that the total displacement at time t+Δt can be written in incremental form

$U^{t+\Delta t} = U^{t} + U^{\Delta t}$(Eq.17)

Accordingly, the strain at time t+Δt is written in the form

$$\varepsilon_{xx}^{t+\Delta t} = \frac{\partial U^{t+\Delta t}}{\partial x} + 1/2\left(\frac{\partial^{t+\Delta t} U}{\partial x}\right)^2 \; 18$$

Which is decomposed in the form

$$\varepsilon_{xx}^{t+\Delta t} = \varepsilon_{xx}^{t} + \varepsilon_{xx}^{\Delta t} \tag{19}$$

in which

$$\varepsilon_{xx}^{t} = \frac{\partial^{t} U}{\partial x} + \frac{1}{2}\left(\frac{\partial^{t} U}{\partial x}\right)^2 \tag{20}$$

and

$$\varepsilon_{xx}^{\Delta t} = \frac{\partial^{\Delta t} U}{\partial x} + \left(\frac{\partial^{t} U}{\partial x}\right)\left(\frac{\partial^{\Delta t} U}{\partial x}\right) = \left(1 + \frac{\partial^{t} U}{\partial x}\right)\left(\frac{\partial^{\Delta t} U}{\partial x}\right) \tag{21}$$

The form in Equation (21) is expressed in finite element format by writing

$$U^{t} = N_1 U_1^{t} + N_2 U_2^{t} \tag{22}$$

and

$$U^{\Delta t} = N_1 U_1^{\Delta t} + N_2 U_2^{\Delta t} \tag{23}$$

The form in Equations (21) and (23) gives

$$\varepsilon_{xx}^{\Delta t} = \left(1 + N_{1,x} U_1^{t} + N_{2,x} U_2^{t}\right)\left(N_{1,x} U_1^{\Delta t} + N_{2,x} U_2^{\Delta t}\right) \tag{24}$$

which, if Equation (9) is plugged into it, gives

$\varepsilon_{xx}^{\Delta t} = \left[B^{t}\right]\left\{U\right\}$(Eq.25)

Where,

$$\left[B^t\right] = \left(1 + \frac{U_2^t - U_1^t}{L}\right)\left[-\frac{1}{L} + \frac{1}{L}\right] \tag{26}$$

Inserting the form in Equation (26) into Equation (12) produces the element stiffness matrix at time *t* as

$$\left[K^t\right] = \left(1 + \frac{U_2^t - U_1^t}{L}\right)\begin{bmatrix} EA/L & -EA/L \\ -EA/L & EA/L \end{bmatrix} \tag{27}$$

which can be used to analyze nonlinear structural systems solved by the Newton-Raphson method and modified Newton-Raphson. In the incremental approach for structural problems with finite displacements, modified Newton-Raphson iterations enhance convergence efficiency and stability by updating the stiffness matrix only when significant changes occur. Each iteration adjusts displacement based on residual forces, recalculating internal forces incrementally. This balances accuracy and computational efficiency.

Modified Newton Raphson Method

The basic Equation is solved in nonlinear analysis at time t+Δt,

$$R^{t+\Delta t} - F^{t+\Delta t} = 0 \tag{28}$$

Since $F^{t+\Delta t}$ depends nonlinearly on the displacement of the nodal points, the solution must be iterated. Modified Newton-Raphson iteration produces

$$\Delta R^{(i-1)} = {}^{t+\Delta t}R - {}^{t+\Delta t}F^{(i-1)} \tag{29}$$

$$K^t \Delta U^{(i)} = \Delta R^{(i-1)} \tag{30}$$

t+Δt,

$$U^{(i)} = {}^{t+\Delta t}U^{(i-1)} + \Delta U^{(i)} \tag{31}$$

with

$^{t+\Delta t,}U^{(0)} = U^t ;\ ^{t+\Delta t}F^{(0)} = F^t$

Equation (29) calculates the unbalanced load vector, producing an incremental displacement via Equation (30). According to Figure 17, instability in the structure emerges at β values around 0.10, posing risks like structural failure. A β value near 0.10 indicates an increasing imbalance in load distributions, potentially causing localized overstresses or buckling. Instability may lead to sudden displacements, urging thorough analysis and structural adjustments to maintain system safety and integrity. Iterations continue until the unbalanced load vector $\Delta R^{(i-1)}$ or incremental displacement $\Delta U^{(i)}$ is small enough. The modified Newton Raphson iteration requires fewer stiffness adjustments than the complete Newton Raphson iteration. The critical difference between modified and full Newton-Raphson iterations is how often the Jacobian matrix is updated. The complete method updates it every iteration for rapid convergence but is computationally intensive. The modified method updates it less frequently, reducing computational effort but potentially needing more converging iterations. The modified Newton-Raphson iteration adjusts stiffness less regularly than the complete Newton-Raphson method, updating it only when significant changes occur in the structure's response. This approach reduces adjustments, improving convergence and reducing computational effort, particularly in gradual response changes. It modifies the stiffness matrix using an agreed-upon equilibrium configuration. The agreed-upon equilibrium configuration is essential when modifying the stiffness matrix as a reference for adjustments due to changes in material properties, geometry, or external influences. This ensures modifications align with the structure's actual behavior under load, including non-linearities and inelastic responses.

Consequently, accurate recalibration maintains structural integrity and predictability, enabling precise simulations and analyses under various load conditions. The timing for updating the stiffness matrix is determined by the extent of nonlinearity in the system's response. As the reaction grows more nonlinear, the frequency of updates must rise accordingly.

Schematically, the iteration for each loading interval runs as follows.

1. 1st iteration, calculate $^tK^{(0)}=K_L^{(0)}$; with $R^{(0)}= R_i$ and $F^{(0)}=0$; from Equation (30) it is solved to get $U^{(0)}$; $U^t = U^{(0)}$
2. 2nd iteration, calculate $F^{(i)}$ with $R^{(0)}= R_i$ and $^tK^{(0)}= K_L^{(0)}$; solve Equation (30) to get $U^{(1)}$; $U^{t+\Delta t} = U^t + U^{(n)}$
3. Iteration n+1, calculate F_n with $R^{(n)}= R^{(i)}$ update K, namely $K_n^i = K_L^i + K_{NL}^i$ Solve Equation (30) to get $U^{(n)}$; $U^{t+\Delta t} = U^t + U^{(n)}$

Figure 3. Modified Newton Raphson method

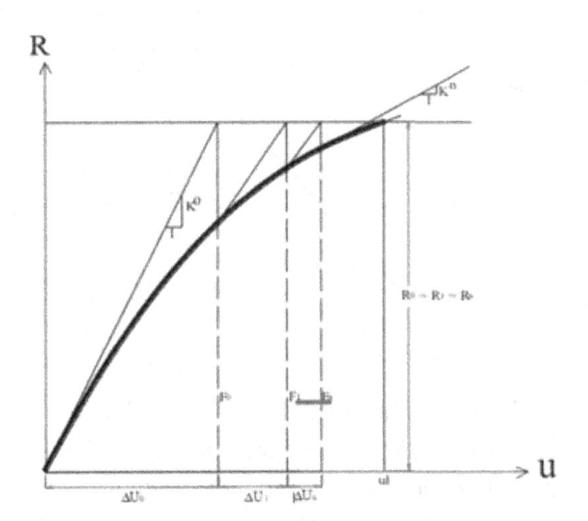

The iteration continues until it converges. The n value mentioned above can be used in various ways, such as 4, 5, or 6. The convergence criterion can be determined by comparing the remaining forces with the square root of the total displacement. The iteration mechanism is shown graphically in Figure 3.

Equation (29) calculates the unbalanced load vector, producing an incremental displacement via Equation (30). Iterations continue until the unbalanced load vector $\Delta R(i-1)$ or incremental displacement $\Delta U(i)$ is small enough.

Jack Method

As explained at the beginning of the chapter, the jack method is used to obtain a snap-through curve from the P-U curve for nonlinear structural response. This jack system solution means increasing structural stiffness due to adding pendel elements to the structural system with the pendel elements' stiffness. More details are shown in Figure 4. Incorporating pendel elements via the Jack Method enhances stiffness and stability systematically. Engineers identify critical points and place pendel elements strategically to supplement support against lateral or torsional loads. Incremental load application and observation enable optimization for effectively withstanding anticipated loading conditions.

(a) Effect of Jack (b) Jack (c) Minus Jack

Figure 4. Jacking method

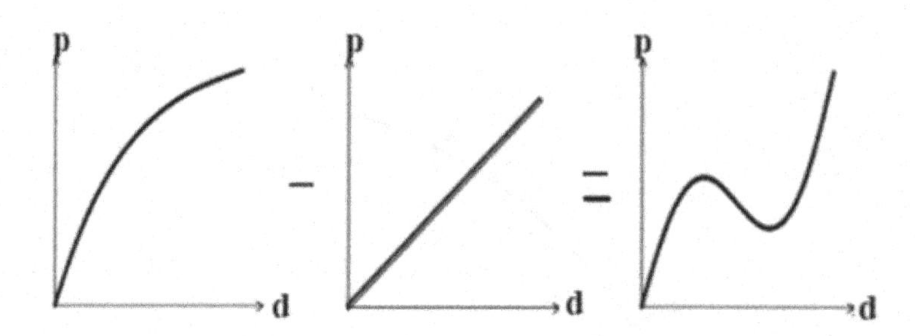

In Figure 4, it can be seen that the addition of pendel elements in the analysis increases stiffness. So, to get the actual structural behavior, the behavior of the structural system due to the addition of pendel elements is reduced by the stiffness of the pendel elements themselves, as in Figure 4. This results in structural behavior, as in Figure 4.

Computer Programming

The structural analysis that has been presented will be compiled into a computer package program. Programming is carried out using computer software with the high-level language Fortran, which is most often applied in engineering disciplines. Fortran excels in advanced structural analysis with high numerical computing performance and efficient large-scale computation handling. Its array-oriented syntax and support for matrix/vector operations streamline finite element methods, while robust libraries and parallel processing enhance efficiency. Fortran's extensive engineering use provides a reliable foundation with proven algorithms and legacy code for optimizing structural analysis programs. The program is created in a modular and structured manner to make it easy to develop. The program consists of principal and subroutine programs that branch out to form a hierarchical structure. The main program contains the primary sequence of all the required work; in other words, it is the highest in the program structure because it functions as the program subroutine coordinator.

Meanwhile, a subroutine is a group of calculations that has a calculation process whose function can be called repeatedly in the main program. This makes it easier to trace if an error occurs; compiling and combining modules (linking) becomes much more efficient. Translating from the source program (source code) into machine language requires an interpreter known as a compiler, which functions to compile the program (compile program). The algorithm that has been prepared is poured into the source program and then compiled. If there are no errors (bugs) in the compilation process, the program build process continues to produce a program application file with the extension .exe. Creating a .exe program file offers significant benefits, especially in research requiring robust software. As a standard Windows executable, it ensures easy distribution, installation, and optimal integration with system resources for enhanced performance. Additionally, .exe files support essential security features like code signing, boosting the application's trust and reliability.

Program Validation

The program that has been prepared and named NLG-STAP will be tested for correctness by carrying out a validation process. This validation process uses a plane joint frame structure consisting of two pendel elements and three vertices, as in Figure 5. The input data shown in Figure 6 is entered by reading a file in the format shown in Table 1.

Figure 5. Modeling of joint frame structure

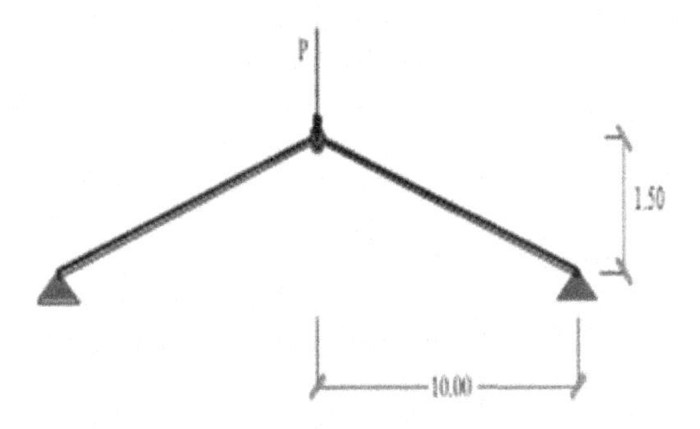

Table 1. Input data format

S.No	Data Block Name	Number of Rows
1	Project title	1
2	Structure title	1
3	General parameters	1
4	Custom parameters	1
5	element type, incidence, area	NEL
6	Point coordinates	NOD
7	Restraint	ICOs

continued on following page

Table 1. Continued

S.No	Data Block Name	Number of Rows
8	Encumbrance title	1
9	Load parameters	1
10	Point load	INL

Figure 6. Program input data

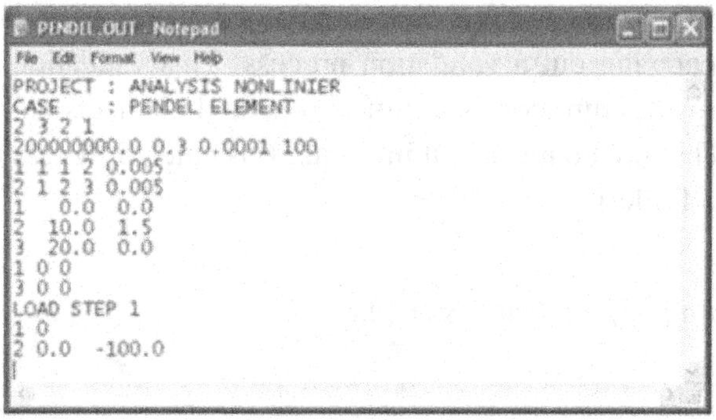

```
PENDEL.OUT - Notepad
File Edit Format View Help
PROJECT : ANALYSIS NONLINIER
CASE    : PENDEL ELEMENT
2 3 2 1
200000000.0 0.3 0.0001 100
1 1 1 2 0.005
2 1 2 3 0.005
1   0.0  0.0
2  10.0  1.5
3  20.0  0.0
1 0 0
3 0 0
LOAD STEP 1
1 0
2 0.0  -100.0
```

Figure 7. Program output data

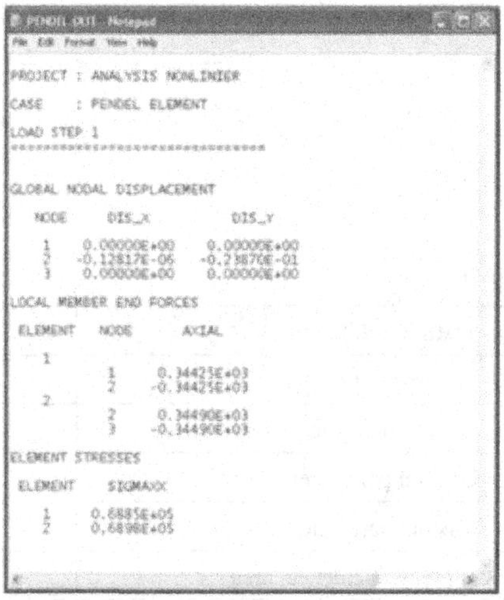

```
PENDEL.OUT - Notepad
File Edit Format View Help

PROJECT : ANALYSIS NONLINIER

CASE    : PENDEL ELEMENT

LOAD STEP 1
********************************

GLOBAL NODAL DISPLACEMENT

   NODE      DIS_X           DIS_Y

    1    0.00000E+00    0.00000E+00
    2   -0.12817E-06   -0.23870E-01
    3    0.00000E+00    0.00000E+00

LOCAL MEMBER END FORCES

 ELEMENT   NODE       AXIAL

    1
            1     0.34425E+03
            2    -0.34425E+03
    2
            2     0.34490E+03
            3    -0.34490E+03

ELEMENT STRESSES

 ELEMENT     SIGMAXX

    1     0.6885E+05
    2     0.6898E+05
```

The output data shown in Figure 7 is then compared with the same structural analysis output data obtained using the SAP (Structural Analysis Program) package program presented in Table 1.

Table 2. Comparison of program output results

S.No.	Comparison	Program	SAP
1	Y Direction Shift at Node 2	-0.0239	-0.02751
2	Axial Internal Force	346.25	346.328
3	Axial Stress	67950	68750

The results obtained using the program that has been prepared are close to the run results using the SAP package program. The displacement in the Y direction at nodal point 2 is 13.2% different from the SAP output. This is because the program incorporates geometric nonlinearities, whereas SAP is based on elastic linear analysis. The Jack Method addresses geometric nonlinearities by considering incremental deformations from applied loads, impacting the structural response's geometry. Unlike SAP's linear analysis, assuming small displacements and linear material behavior, the Jack Method allows for large deformations and considers geometric nonlinearity. Incremental loading and observation enable capturing complex interactions, accurately representing structural behavior under nonlinear conditions. However, the axial stress is the same because the frame structure being analyzed is a static system where the reaction force is not affected by displacement.

RESULTS AND DISCUSSION

As discussed previously, the structural analysis has been outlined in a computer package using the finite element method for nonlinear analysis.

The case study is limited to the case of joint frame structural systems. As shown in Figure 8.

Figure 8. Type I Truss structure

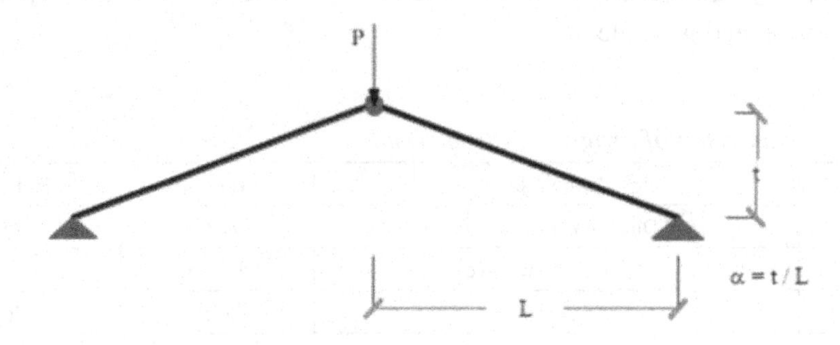

A joint frame structure with two-rod elements is loaded vertically by a concentrated load at the node point in the middle of the structure. The loading is carried out several times until the structure experiences nonlinear behavior. The discussion includes the influence of geometric nonlinear assumptions on structural behavior when,

1. variation in the ratio of the span of the structure L to the height of the structure T, from now on referred to as α,

Variations in modulus of elasticity E and cross-sectional area A, from now on referred to as β. Parameters *a* and *β* are pivotal in structural analysis, shaping material or structural behavior under load. *a* denotes material properties like stiffness, affecting deformation resistance and stability, while *β* represents geometric or load-related factors influencing stress distribution. Accurate integration of *a* and *β* values into analysis ensures dependable structural design by predicting responses, deformation patterns, and potential failure modes. The output results are images of load vs. vertical displacement of the structure-load center for each variation of α and β. The linearity of elastic systems allows engineers to predict structural behavior under various loads using the principle of superposition, where the total response is the sum of individual responses. This simplifies analyses and ensures that stress, strain, and displacement are directly proportional to applied loads. Predictability aids in designing safe, efficient structures by enabling accurate assessments without extensive computational or experimental efforts.

The output results are images of load vs. vertical displacement of the structure-load for each variation of α and β.

High Structure Problems

In this case, large values are applied to $0.30 \leq \alpha$ £ 0.50. The analysis is carried out for type I structures, as seen in Figure 8 above. Comparisons will be made for positive and negative variations.

Comparison of Calculation Results for α Varies Positively

The first case study was carried out for type I structures for $0.30 \leq \alpha \leq 0.50$ with a multiple of α=0.05. The magnitude of β is the same. The loading stage will be carried out 40 times.

Figure 9. Load vs. displacement for $0.30 \leq \alpha \leq 0.50$

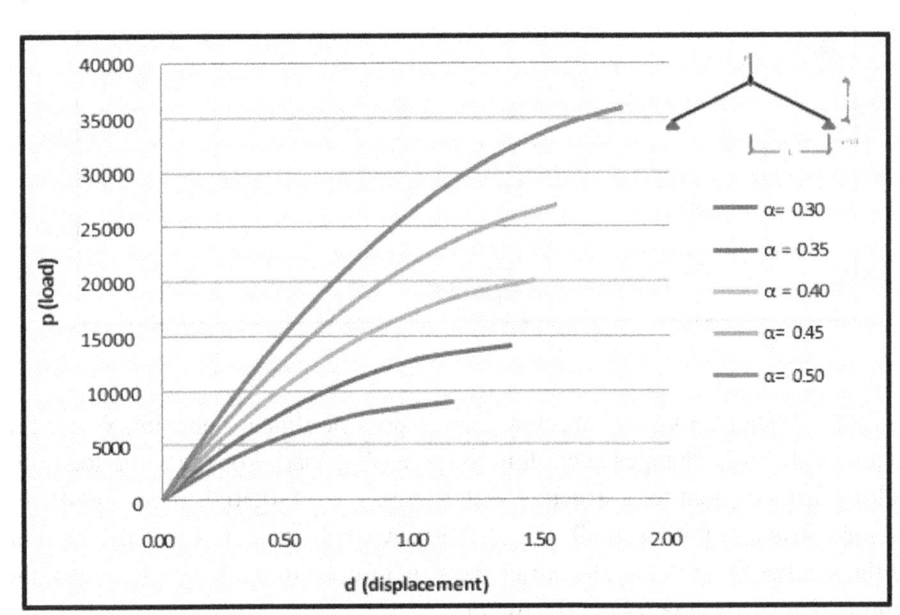

Figure 9 shows that the structure experiences nonlinear behavior, where the stiffness changes according to increasing load. For the same load, the resulting displacement for a structure with a larger α ($\alpha = 0.50$) is smaller than for a structure with a small α (a = 0.30). The greater α, the greater the maximum load that the structure can support. Meanwhile, the stiffness of the structure will decrease as the structure increases.

Comparison of Calculation Results for α Varies Negatively

The second case study was carried out for type I structures, but for $-0.30 \leq \alpha \leq -0.50$ with a multiple of a=0.05. The magnitude of β is the same. The loading stage will be carried out 40 times.

Figure 10. Load vs. Displacement For $-0.30 \leq \alpha \leq -0.50$

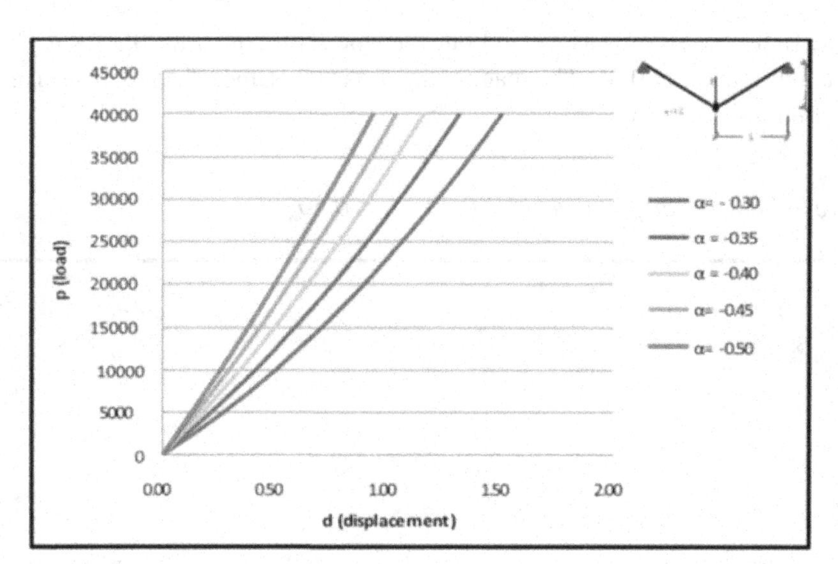

Figure 10 shows that the structure experiences nonlinear behavior, where the structure's stiffness changes according to increasing load. For the same load, the resulting displacement for a structure with a larger α (a = -0.30) is more significant than for a structure with a small α (a = -0.50). Apart from that, it can also be seen that the smaller the α value, the stiffer the structure, so that for the same load, the resulting deformation will be more minor.

Comparison of Calculation Results for Positive and Negative Varying α

The third case study compares the calculation results for the exact value of α with opposite signs, the case for α = -0.50 and α = 0.50 for type I structures.

Figure 11. Load vs. displacement for α = -0.50 and α = 0.50

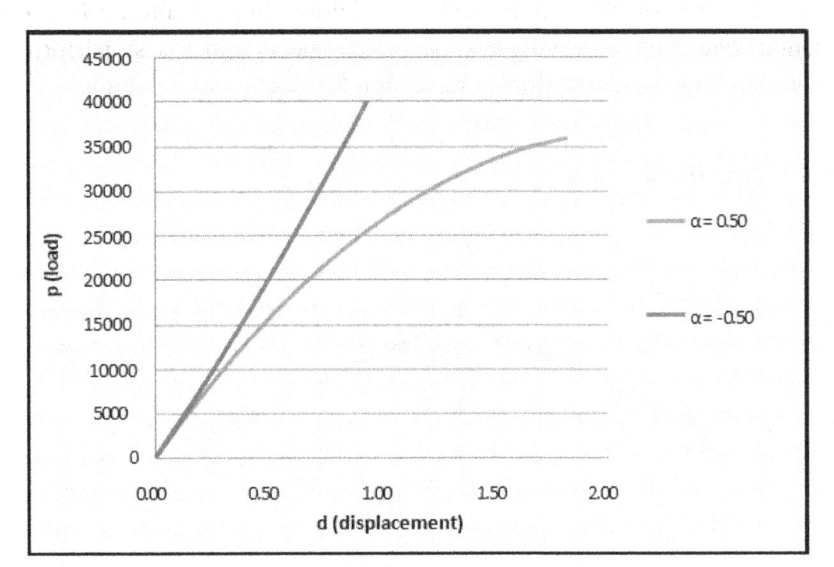

Figure 11 shows that the structure experiences nonlinear behavior, where the structure's stiffness changes according to increasing load. For a, which has a positive value in this case (a = 0.50), the structure experiences softening behavior where the structure's stiffness decreases as the displacement increases. Meanwhile, for α with a value of (α = - 0.50), the structure experiences stiffening behavior where the stiffness of the structure actually increases with increasing displacement. For this reason, it can be seen that for the same load, the resulting structural displacement for a structure with a positive alpha value is greater than for a structure with a negative alpha value.

Snap-Through Problem

Nonlinear structural behavior when peak stress occurs, namely at the critical point, the structure will experience snap-through behavior. The modified Newton-Raphson method adeptly addresses nonlinear structural challenges such as softening, stiffening, and snap-through phenomena. Its iterative updates accurately capture complexities and instabilities, ensuring dependable analysis. With its adaptability to stiffness and geometry changes, it serves as a robust tool for nonlinear structural analysis across varied loading conditions. This happens for small values of α. For this reason, the case of small α joint frame structures has been explained in the previous subchapter. given additional structure as seen in Figure 12. The structural system analyzed is a joint frame structure with interconnected beams and columns designed to resist various loads. The focus is on joint behavior under applied loads, particularly their capacity to transfer forces without significant deformation or failure. By applying incremental loads and observing responses, engineers can assess performance, identify failure modes, and optimize the design for safety and reliability.

Figure 12. Type II Truss Structure

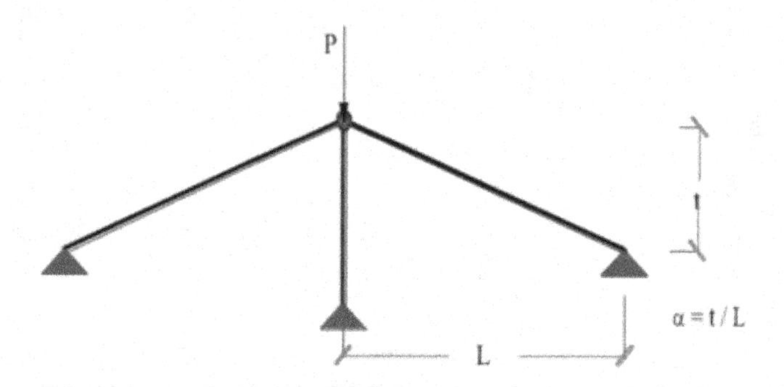

Comparison of Calculation Results for α Varies Positively

The fourth case study was carried out for type II structures for $0.05 \leq \alpha \leq 0.25$ with a multiple of α=0.05. Where for each α a variation of $0.00 \leq \beta \leq 0.50$ will be applied, so that the β value that is suitable for each variation of α can be determined. The loading was carried out 30 times.

From Figure 13, it can be seen that the behavior of the structure experiences three nonlinear conditions, namely softening, snap-through and stiffening. For a structure with a large β value in this case ($\beta = 0.50$) to reach the critical point the applied load must be greater than for a structure with a small β value in this case ($\beta = 0.10$. In addition, the application of the same load for a structure with a large β larger results in a smaller displacement than a structure with a smaller β. This is because the greater β, the greater the stiffness of the structure, inversely proportional to the resulting displacement which is actually smaller. For $\alpha = 0.05$ the structure starts to become unstable for $\beta \leq 0.20$.

Figure 13. Load vs. displacement with $0.10 \leq \beta \leq 0.50$ For $\alpha = 0.05$

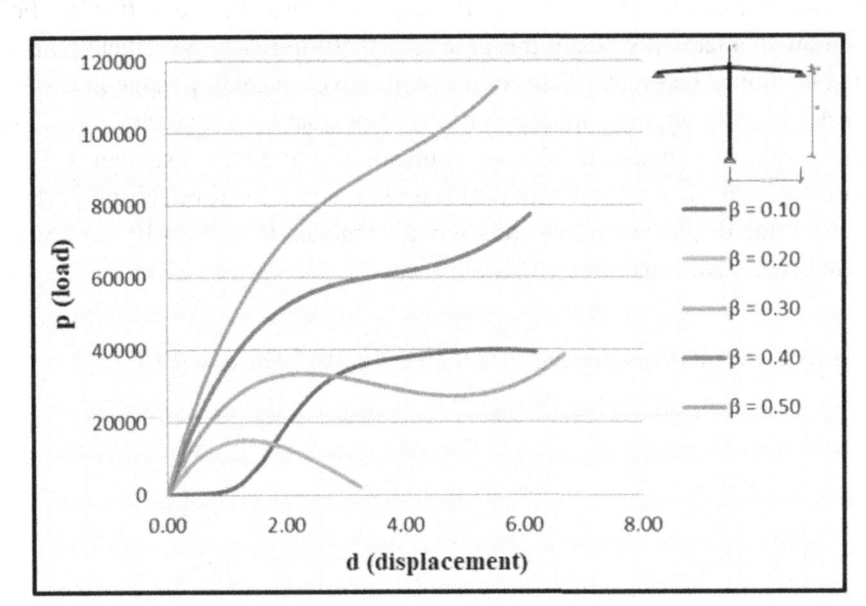

Figure 14. Load vs. displacement with $0.10 \leq \beta \leq 0.50$ For $\alpha = 0.10$

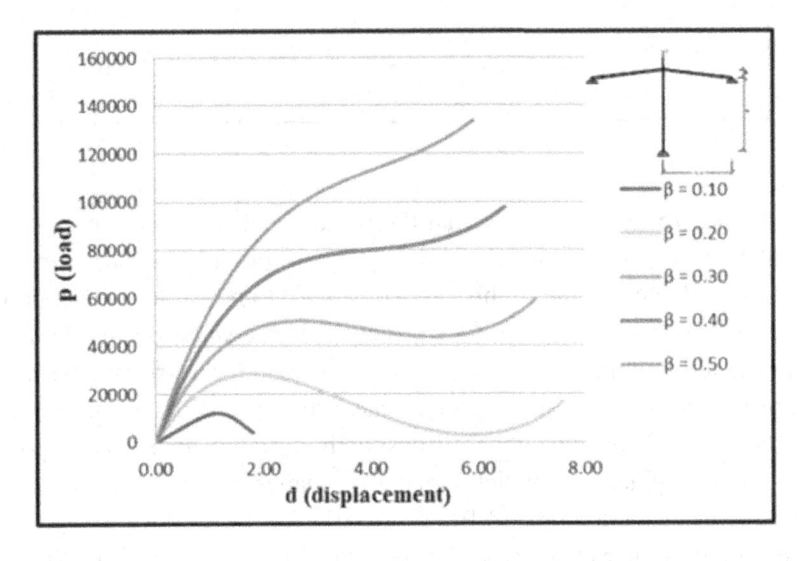

From Figure 14, it can be seen that the behavior of the structure experiences three nonlinear conditions, namely softening, snap-through and stiffening. For a structure with a large β value in this case (β = 0.50) to reach the critical point the applied load must be greater than for a structure with a small β value in this case (β = 0.10. In addition, the application of the same load for a structure with a large β larger results in a smaller displacement than a structure with a smaller β. This is because the greater β, the greater the stiffness of the structure, inversely proportional to the resulting displacement which is actually smaller. For α = 0.10 the structure starts to become unstable for β ≤0.10.

Figure 15. Load vs. displacement with 0.10 ≤ β ≤ 0.50 For α = 0.15

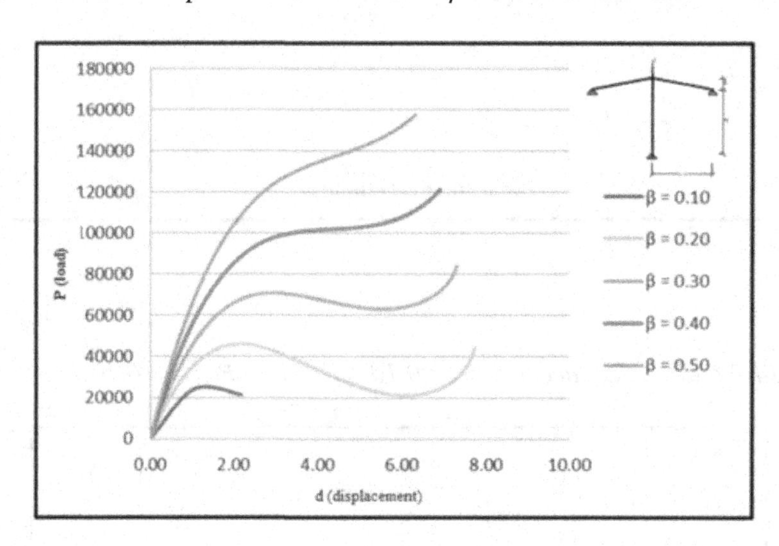

From Figure 15, it can be seen that the behavior of the structure experiences three nonlinear conditions, namely softening, snap-through and stiffening. For a structure with a large β value in this case (β = 0.50) to reach the critical point the applied load must be greater than for a structure with a small β value in this case (β = 0.10. In addition, the application of the same load for a structure with a large β larger results in a smaller displacement than a structure with a smaller β. This is because the greater β, the greater the stiffness of the structure, inversely proportional to the resulting displacement which is actually smaller. For α = 0.15 the structure starts to become unstable for β ≤0.10.

From Figure 16, it can be seen that the behavior of the structure experiences three nonlinear conditions, namely softening, snap-through and stiffening. For a structure with a large β value in this case (β = 0.50) to reach the critical point the applied load must be greater than for a structure with a small β value in this case (β = 0.10. In addition, the application of the same load for a structure with a large β A larger displacement results in a smaller displacement than a structure with a smaller β. This is because the greater β, the greater the stiffness of the structure, inversely proportional to the resulting displacement, which is actually smaller. For α = 0.20, the structure starts to become unstable for β ≤0.10.

Figure 16. Load vs. displacement with $0.10 \leq \beta \leq 0.50$ For $\alpha = 0.20$

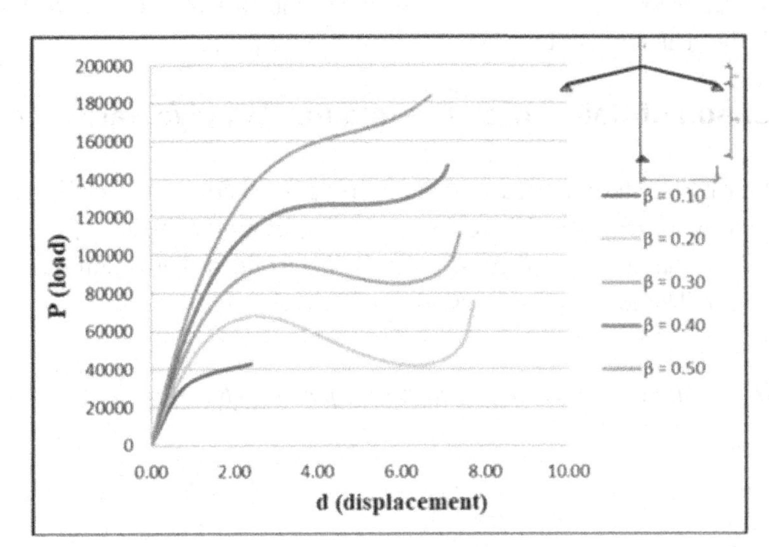

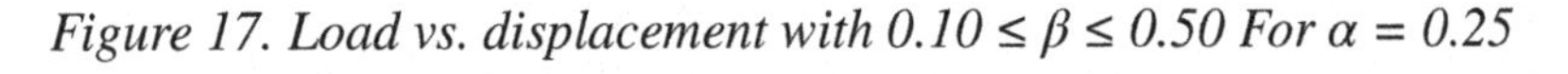
Figure 17. Load vs. displacement with $0.10 \leq \beta \leq 0.50$ For $\alpha = 0.25$

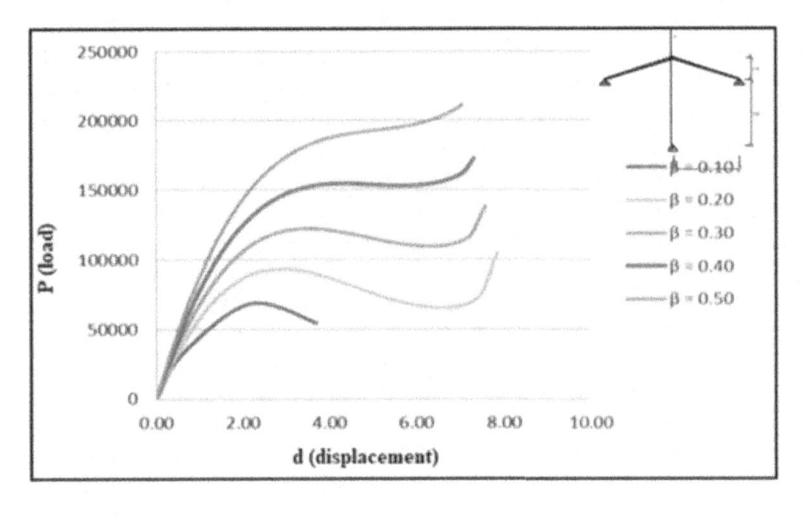

From Figure 17, it can be seen that the behavior of the structure experiences three nonlinear conditions, namely softening, snap-through and stiffening. For a structure with a large β value in this case (β = 0.50) to reach the critical point the applied load must be greater than for a structure with a small β value in this case (β = 0.10. In addition, the application of the same load for a structure with a large β A larger displacement results in a smaller displacement than a structure with a smaller β. This is because the greater β, the greater the stiffness of the structure, inversely proportional to the resulting displacement, which is actually smaller. For α = 0.25, the structure starts to become unstable for β ≤0.10.

Comparison of Calculation Results for Negative Varying α

The fourth case study was carried out for type II structures for -0.05 ≤ α ≤ -0.25 with a multiple of α=0.05. Where for each α a variation of 0.00 ≤ β ≤ 0.50 will be applied, so that a value range of β can be determined that is suitable for each variation of α. The loading was carried out 30 times.

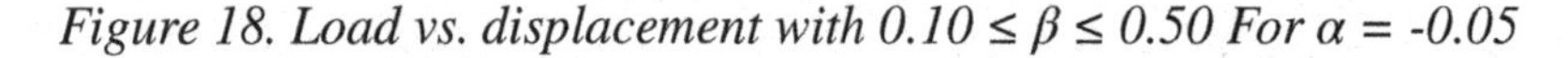

Figure 18. Load vs. displacement with 0.10 ≤ β ≤ 0.50 For α = -0.05

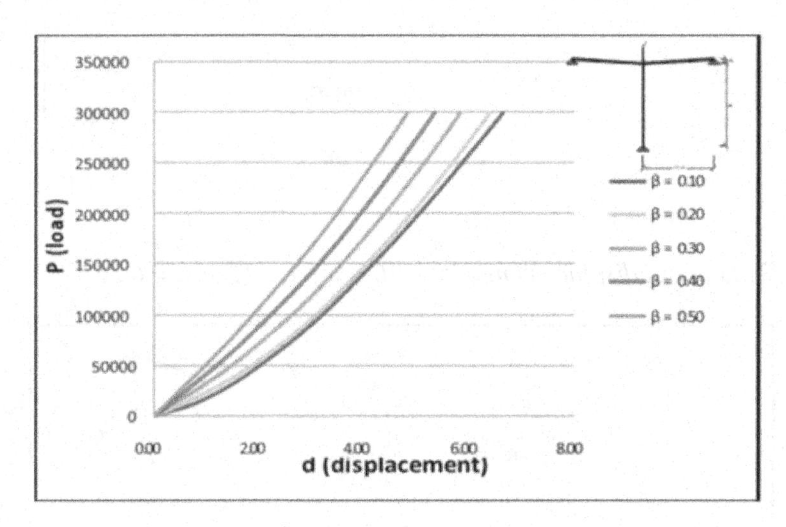

Figure 19. Load vs. displacement with $0.10 \leq \beta \leq 0.50$ *For* $\alpha = -0.10$

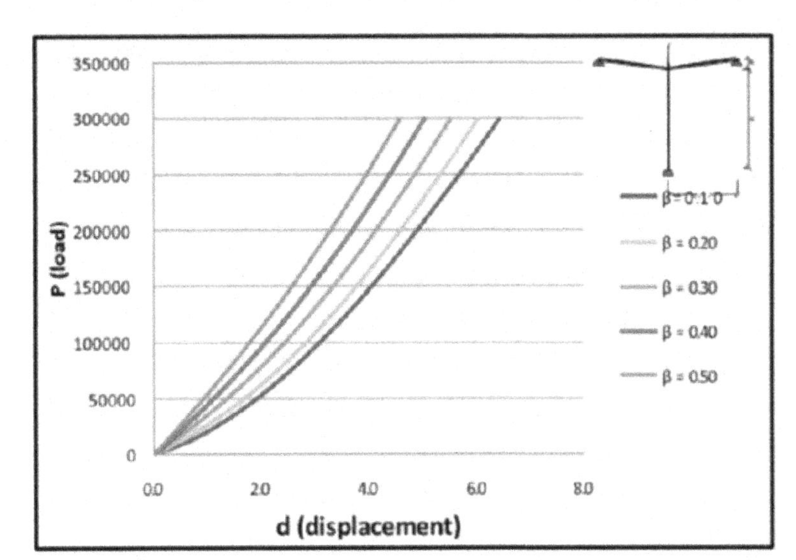

Figure 20. Load vs. displacement with $0.10 \leq \beta \leq 0.50$ *For* $\alpha = -0.15$

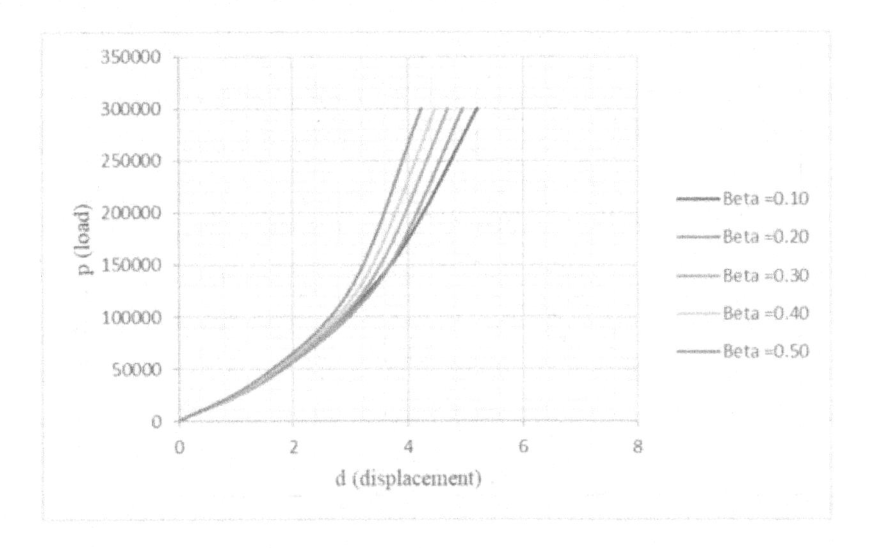

Figure 21. Load vs. displacement with 0.10 ≤ β ≤ 0.50 For α = -0.20

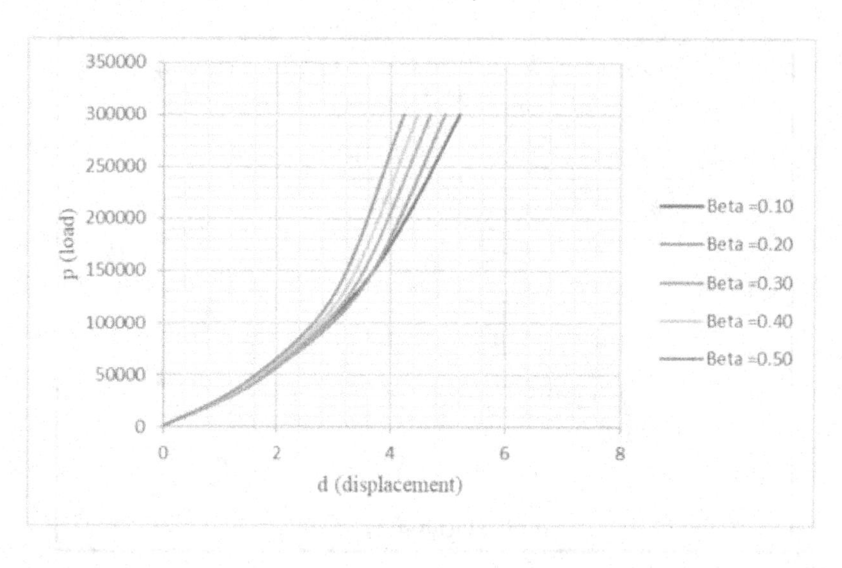

Figure 22. Load vs. displacement [With 0.10 ≤ β ≤ 0.50 for α = -0.25]

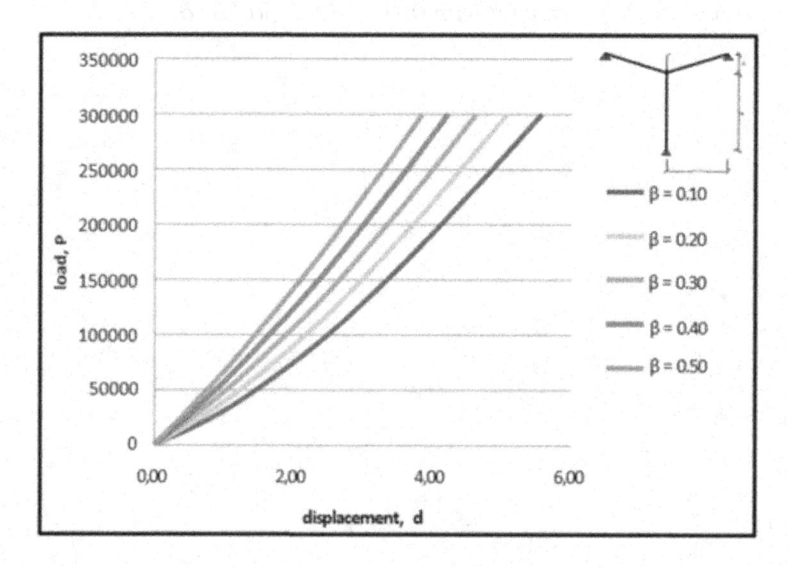

From Figure 18 to Figure 22, it can be seen that the behavior of the structure only experiences stiffening conditions. The stiffness of the structure actually increases with increasing displacement. Applying the same load to a structure with a smaller β results in a greater displacement than a structure with a larger β.

Comparison of Calculation Results for Positive and Negative Varying α

Figure 23. Load vs. displacement for α = -0.2 and α = 0.2

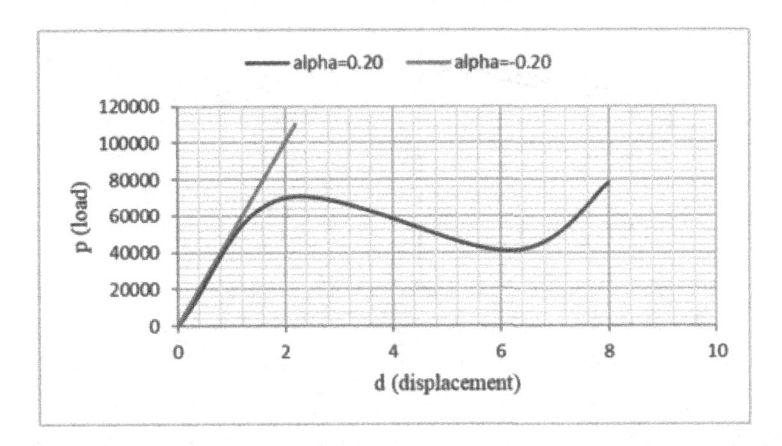

The fifth case study is to compare the calculation results for the same value of α with opposite signs, the case for α = -0.20 and α = 0.20 for type II structures.

From Figure 23, it can be seen that the same β and different α values produce different behavior. For α =0.2 the structure has three conditions, softening, snap-through and stiffening. Meanwhile, for α = -0.2, the structure only experiences a stiffening condition. The stiffness of the structure actually increases with increasing displacement.

Case of Jointed Skeletal Systems

Figure 24. Modelling of joint frame structure

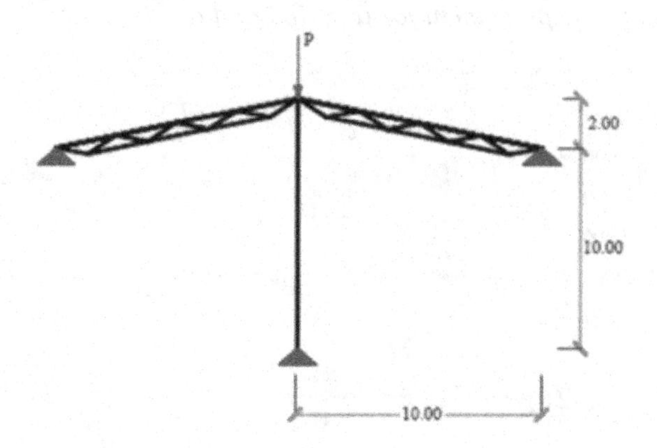

As a further application of the program for Joint Frame Structures, it is applied to the structure in Figure 24. The structure experiences a concentrated load in the middle of the structure span. The loading was carried out 40 times. The program output results for this structure are presented in Figure 25. which is compared with the pendel structure with varying height to span ratios, namely $\alpha = 0.2$.

Joint frame structures are stiffer than pendel structures due to their loading response mechanisms. Interconnected beams and columns in joint frames resist deflections under loads. Pendel structures, with flexible connections, require additional stiffness for minimal deformation under static loads, making joint frames preferable for stability in high-rise buildings or bridges.

Figure 25. Load vs. displacement for α = 0.2

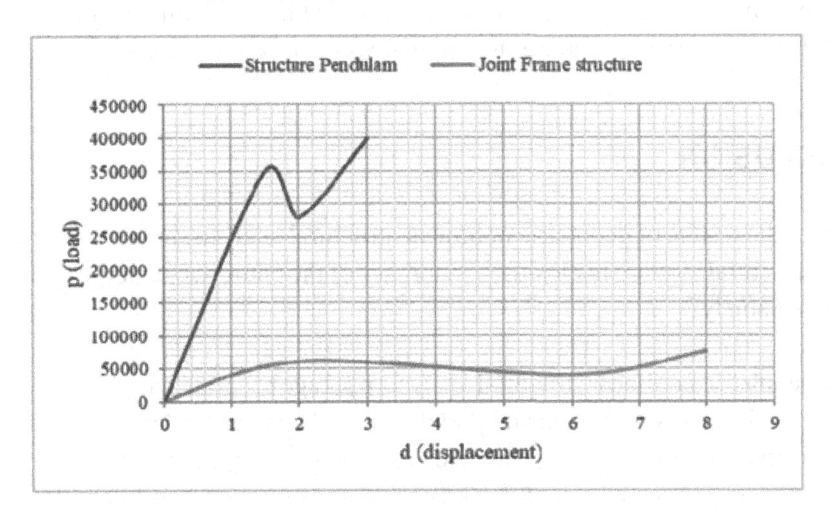

In Figure 25, it can be seen that the joint frame structure experiences softening, snap-through and stiffening conditions. Apart from that, it can also be seen that the stiffness of the joint frame structure is stiffer than that of the pendel structure. In the joint frame structure the maximum load that the structure can carry is 351892.395 KN which gives a displacement of the structure in the middle of the span of 1.42 m, while in the pendel structure the maximum load that the structure can carry is 67626.903 KN which gives a displacement of the structure in the middle of the span of 2.36 m.

Discussion of Results

From all the results that have been obtained previously, it can be concluded that the application of the Newton-Raphson method can only be used for structural systems with a positive stiffness determinant. The iterative modified Newton-Raphson method for solving nonlinear structural problems updates the solution until convergence. It begins with an estimate, evaluates residual forces, and computes corrections using linearized equations, capturing nonlinear behavior for accurate convergence. Understanding snap-through phenomena in structural systems is crucial for addressing sudden instabilities, involving rapid transitions between stable states and significant deformation. By studying snap-through behavior, engineers can identify vulnerabilities, develop predictive models, and make design modifications to enhance structural robustness, especially in aerospace and civil engineering applications. To determine a stiffness matrix with a negative value, other methods must be used, such as the arc-length method or the jack method. The stiffness of the jack needs to be chosen so that its use is optimal for the case at hand; in other words, for the α at hand it is necessary to choose the optimal value of β.

CONCLUSION

From the results of geometric nonlinear structural analysis, in this case the joint frame structural system. In closing, a summary of the entire scope of the discussion is given, namely:

1. A geometric nonlinear structural analysis package program based on the principle of virtual displacement and prepared based on a finite element formulation has been completed. The principle of virtual displacement offers a flexible framework for deriving equilibrium equations, accommodating complex loads, materials, and geometries. It simplifies finite element formulations and effectively handles both linear and nonlinear problems. This versatility ensures accurate and efficient solutions, especially for large deformations or nonlinear material behavior.
2. The program has been compiled, checked for correctness, both in syntax and execution logic, and tested and applied to several joint frame structural systems.
3. The program is arranged in modules so that it can be modified quickly according to needs.
4. The program has been validated by comparing the results obtained with the results obtained using existing package programs available in the field, in this case the SAP package program. Validation shows quite close results.

After testing the joint frame system which was applied in several case examples, the conclusions regarding the results achieved include:

1. When using this program, the addition of loading steps must be carefully considered and arranged to obtain displacements that are in accordance with the building blocks behavior that occurs.
2. Losses in both elastic modulus and cross-sectional area make nonlinear behavior more pronounced. Greater displacement happens for the same load when the material is softer.
3. The impact of the ratio of a structure's height to its span makes nonlinear behavior more apparent as well. Geometric nonlinearity becomes more apparent with decreasing comparison sizes.

4. Softening, snap-through, and stiffening are the three structural responses that the program can handle.

REFERENCES

Asiedu, Chapman-Wardy, & Doku-Amponsah. (2021). *A Modification of Newton Method for Solving Non-linear Equations.* Academic Press.

Cacciola, P., & Tombari, A. (2021). Steady state harmonic response of nonlinear soil-structure interaction problems through the Preisach formalism. *Soil Dynamics and Earthquake Engineering*, 144, 106669. 10.1016/j.soildyn.2021.106669

Chegeni, Sharbatdar, & Mahjoub, &Raftari. (2022). *Numerical Methods in Civil Engineering A novel method for detecting structural damage based on data-driven and similarity-based techniques under environmental and operational changes.* Numerical Methods in Civil Engineering. 10.52547/nmce.6.4.16

Fei, Y.-F., Tian, Y., Huang, Y., & Lu, X. (2022). Influence of damping models on dynamic analyses of a base-isolated composite structure under earthquakes and environmental vibrations. Gong Cheng Li Xue. *Gongcheng Lixue*, 39, 201–211. 10.6052/j.issn.1000-4750.2021.07.0500

Jagota, V., Sethi, A., & Kumar, D.-K. (2013). Finite Element Method: An Overview. *Walailak Journal of Science and Technology*, 10, 1–8. 10.2004/wjst.v10i1.499

Kythe, P., Wei, D., & Okrouhlik, M. (2004). An Introduction to Linear and Non-linear Finite Element Analysis: A Computational Approach. Applied Mechanics Reviews -. *Applied Mechanics Reviews*, 57(5), B25. Advance online publication. 10.1115/1.1818688

Langlois & Deville. (2014). *Introduction to the Finite Element Method.* Springer. .10.1007/978-3-319-03835-3_10

Li, G., & Yu, D.-H. (2018). Efficient Inelasticity-Separated Finite-Element Method for Material Nonlinearity Analysis. *Journal of Engineering Mechanics*, 144(4), 04018008. 10.1061/(ASCE)EM.1943-7889.0001426

Li, K., Jarrar, F., Sheikh-Ahmad, J., & Ozturk, F. (2017). Using coupled Eulerian Lagrangian formulation for accurate modeling of the friction stir welding process. *Procedia Engineering*, 207, 574–579. 10.1016/j.proeng.2017.10.1023

Long, H., Wang, Z., Zhang, C., Zhuang, H., Chen, W., & Peng, C. (2021). Nonlinear study on the structure-soil-structure interaction of seismic response among high-rise buildings. *Engineering Structures*, 242, 112550. 10.1016/j.engstruct.2021.112550

Nagarajan, P. (2018). *Matrix Methods of Structural Analysis* (1st ed.). CRC Press. 10.1201/9781351210324

Pepper, D. (2005). *The Finite Element Method: Basic Concepts and Applications*. Taylor & Francis. .10.1201/9780203942352

Pradhan & Snehashish. (2019). Finite Element Method. In *Computational Structural Mechanics*. Academic Press.

Zhang. (2023). Application of finite element analysis in structural analysis and computer simulation. *Applied Mathematics and Nonlinear Sciences*, *9*.10.2478/amns.2023.1.00273

Chapter 3
Fundamentals of Numerical

Shihab A. Shawkat
https://orcid.org/0000-0002-9529-2151
University of Samarra, Iraq

M. Lellis Thivagar
Independent Researcher, India

ABSTRACT

This chapter provides an overview of numerical methods, including their fundamental concepts, the most widely used numerical methods, their application in solving differential equations of practical significance in engineering, and the potential and constraints they offer in enhancing our comprehension of the surrounding world. The chapter encompasses a comprehensive exploration of the practical applications of the finite element method in addressing diverse engineering challenges.

INTRODUCTION

Numerical methods are the mathematical techniques that allow the solution to a problem to be expressed in the form of numbers. For many, numerical methods are a branch of applied mathematics, that is, that part of mathematics interested in solving problems that directly or indirectly affect the interests of man. The term "numerical solution" is often used in conjunction with the "analytical solution" of a problem (also called "exact solution"). The difference between both types of solution is substantial. Let us consider as an example the study of the behavior of a physical or humanistic system. This system satisfies general mathematical laws (equations) (called governing equations); for example, differential or algebraic equations or

DOI: 10.4018/979-8-3693-3964-0.ch003

inequalities in which a set of variables and physical parameters intervene. The analytical solution is a mathematical expression that provides all the information about the behavior of the system, for any value of the variables and parameters involved in the governing equations. It is, therefore, the "universal" solution to the problem that the Pythagoreans, Plato and so many other believers in the digitalization of the world aspired to. On the other hand, the numerical solution expresses the behavior of the system in terms of numbers that are obtained by solving the governing equations for specific values of the variables and parameters of the system. As an example, let's pose the problem of finding the equation that expresses all the possible trajectories of a mobile phone that moves in a straight line always starting from the same point. Obviously, the general solution to the problem is $y = kx$, where y is the coordinate of a point on the trajectory, x is the value of the abscissa and k is the slope of the line. This equation is universal and expresses the infinite set of lines that passes through the origin. Giving values to the slope k and the abscissa x we will find the coordinates of the points of the selected trajectories, characterized by the values of x and y.

A classic example of a numerical solution was that obtained by Archimedes for the approximate value of the number π from the division of the circle into polygons, obtained by progressively increasing the number of sides and dividing the perimeter of each polygon by the radius of the circle. Obviously, by increasing the number of sides of each polygon, the precision in the value of the searched number π also increased. Taking polygons inscribed and circumscribed in the circumference with up to 96 sides, Archimedes managed to limit the value of π between 3.14084 and 3.14285 (Bidwell, 1994).

Figure 1. A method that Archimedes employed to get the value of π by utilizing the perimeter of polygons that were both inside and outside of a circle

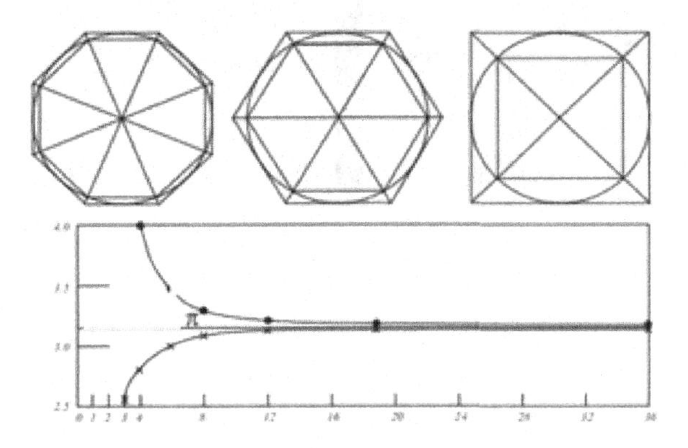

The technique used by Archimedes synthesizes the essence of numerical methods. The numerical solution is obtained by dividing the domain being studied (the circumference) into simple geometric shapes (straight lines), of which all their properties (length) are known. It is also observed that the numerical solution is approximate and improves (converges) by increasing the number of divisions of the domain. Finally, it is very important to note that the "exact" solution to the problem (the exact value of π, one of the incommensurable numbers) is unknown and the numerical solution is the only alternative, as happens in most practical problems.

It is interesting to note that a similar "divide and conquer" technique was used by Chinese calculators in the *5th* century AD. to obtain a dimension of the value of π by dividing the area of the circle into rectangles inscribed and circumscribed to the circumference (Figure 2) (Huang et al., 2021). Numerical methods, therefore, look for numbers, while analytical methods look for mathematical formulas. Obviously, the analytical solution, being universal, contains all the numerical solutions, while from the numerical solution of a problem it is impossible, in general, to deduce the analytical solution.

Figure 2. Approximation of the area of the circle by adding the areas of inscribed and circumscribed rectangles. The black areas indicate the error in the area calculation. This method was used to evaluate the number π by Chinese mathematicians and engineers in the 5th century AD.

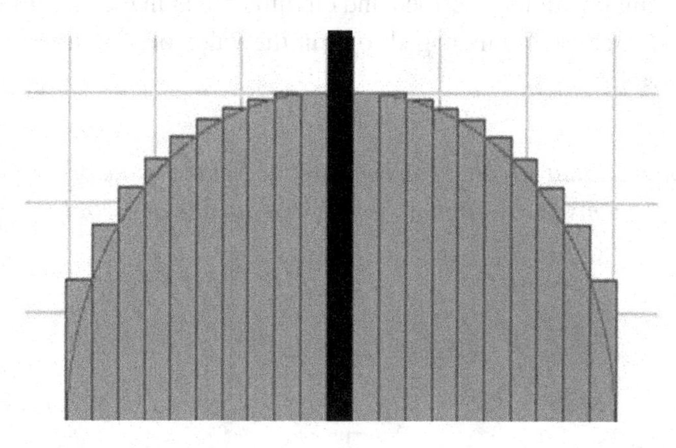

NECESSITY AND LIMITS OF DIFFERENTIAL EQUATIONS

Both numerical and analytical methods share a fundamental starting point: the need to state the problem to be solved in mathematical form. For thousands of years, until the discovery of infinitesimal calculus by Newton and Leibniz, these mathematical forms were variants of polynomial expressions, often intended to solve problems of a geometric nature (such as finding the value of the number π). The revolution in infinitesimal calculus rested fundamentally on its ability to express any natural problem in mathematical form through differential equations. At the same time, Newton, Leibniz and many other distinguished successors (Euler, Cauchy, Lagrange, Navier, Gauss, etc.) laid the foundations for solving these differential equations analytically (Guicciardini et al., 2006). The novel techniques of differential and integral calculus introduced by Newton and Leibniz held a deep significance in the field of mathematics, akin to the enormous impact that the advent of fire had on early humans or the introduction of electricity had on the industrial revolution.

This assertion is not overstated, Prior to the contributions of Newton and Leibniz, there existed a lack of a comprehensive approach to addressing specific physics problems through the formulation of mathematical equations. Examples of such problems include the study of heat propagation within a body, fluid flow, and the equilibrium of an elastic material. Evidently, due to the inability to formulate the problem, its resolution was unattainable. Following the advancements made by Newton and Leibniz, it became feasible to analyses the dynamics of various physical systems, including solids, liquids, and gases, using differential and integral equations. Additionally, there were numerous methods accessible to solve these equations in various scenarios. Although they were typically simplified versions of the broader problem, they facilitated substantial progress in scientific and technical understanding. Thus, while mathematics, as an autonomous science, explored new fields of increasing abstraction, its application to other sciences became increasingly indispensable and effective. This application extended, during the 18th century and the beginning of the 19th, from mechanics and astronomy to the remaining branches of physics; later to all natural sciences and, in the 20th century, to all sectors of knowledge (Imanova, 2022; Tuama et al., 2022).

The initial exhilaration in the scientific community surrounding the first successes of infinitesimal calculus was later dashed by a depressing bit of data. Differential equations could be used to mathematically express any problem; however, the achievement of an "exact" solution to these equations was restricted to certain cases, which sometimes reflected oversimplified versions of reality. The difficulties in finding a general mathematical formula that can be applied to actual problems in science and technology have highlighted the need to investigate alternative methods for solving differential equations. In the early 1900s, scientists observed that discretizing the differential equations for a given problem allowed them to derive the numerical values of the unknown function. These techniques were similar to those used by Archimedes to find the value of the number. The numerical techniques have surfaced (Huang & Multerer, 2022).

The strategy used by all numerical methods is to convert the differential equations governing a given problem into an algebraic system of equations that depend on a finite number of unknowns. Nevertheless, the final system of equations can only be solved with the aid of computers due to the significant amount of unknown variables, which frequently number in the thousands or even millions. This explains why, even though there have been many numerical methods since the 19th century, their substantial development and broad use have accompanied the emergence of contemporary computers in the *20th* century. One way to think of numerical approaches is as the application of numerical values as the main tools for solving a particular problem. In recent decades, the loop that Pythagoras started 2,500 years ago has been successfully closed because numerical methods have shown to be effective in offering conclusive answers to the mysteries of the cosmos.

GENERAL APPROACH OF THE NUMERICAL SOLUTION OF AN ENGINEERING PROBLEM

As mentioned in the previous section, most engineering problems can be expressed through partial differential equations with their corresponding boundary conditions. These equations are usually obtained by balance rules on a differential element within the analysis domain and on its boundaries. Concrete examples of this balance technique are the second-order differential equation that expresses the heat flow balance in a domain Ω (known in its generic form as Poisson's equation), the differential equations of internal equilibrium and equilibrium in the contour of the Ω domain in solid mechanics, the momentum and mass balance equations in fluid dynamics, etc.

In general, these differential equations (1_a,1_b) can be expressed as:

$$Au - b = 0 \text{ in } \Omega \ (1_a)$$

$$Bu - q = 0 \text{ in } \Gamma \ (1_b)$$

In the previous expression *A* and *B* are differential operators that act on the vector of unknowns *u; b* and *q* are vectors containing data from the problem. Operator *A* is obtained from the balance of "flows" over the domain Ω, while operator *B* results from the balance conditions in a domain close to the contour Γ.

As an example of Eqs. (1) Let us consider the Poisson equation that governs the transfer of heat by conduction in a two-dimensional domain:

In the previous expressions φ is the temperature, k_x and k_y are the conductivities of the material in the x and y directions, respectively, Q the heat source on the domain Ω, q the heat lost in the direction normal to the contour Γ_q, with n_x and n_y the directions of the vector normal to said contour and φ the temperature prescribed in the contour $\Gamma\varphi$. The union of the contours Γq and $\Gamma\varphi$ is the total contour Γ of the domain Ω.

From the equations (2) and (3) the operators and vectors of the equations are derived. (1) as:

$$\frac{\partial}{\partial x}\left(k_x \frac{\partial \varphi}{\partial x}\right) + \frac{\partial}{\partial y}\left(k_y \frac{\partial \varphi}{\partial y}\right) + Q = 0 \; in \; \Omega \tag{2}$$

$$k_x n_x \frac{\partial \varphi}{\partial x} + k_y n_y \frac{\partial \varphi}{\partial y} + \overline{q} = 0 \; in \; \Gamma_q \tag{3}$$

$$\varphi - \overline{\varphi} = 0 \; in \; \Gamma_\varphi$$

In the previous expressions φ is the temperature, k_x and k_y are the conductivities of the material in the x and y directions, respectively, Q the heat source on the domain Ω, $\overline{q}$ the heat lost in the direction normal to the contour Γ_q, with n_x and n_y the directions of the vector normal to said contour and φ the temperature prescribed in the contour $\Gamma\varphi$. The union of the contours Γq and $\Gamma\varphi$ is the total contour Γ of the domain Ω.

From the equations (2) and (3) the operators and vectors of the equations are derived. (1) as:

$$A = \left[\frac{\partial}{\partial x} k_x \frac{\partial}{\partial x} + \frac{\partial}{\partial x} k_y \frac{\partial}{\partial y}\right] \tag{4}$$

$$B = \left[k_x n_x \frac{\partial}{\partial x} + k_y n_y \frac{\partial}{\partial x}\right] \tag{5}$$

$$u = \left[\varphi\right] \; b = \left[-Q\right] \; and \; q = \left\{\frac{\overline{q}}{\overline{\varphi}}\right\} \tag{6}$$

In essence, the role of numerical methods is to solve in numerical form the governing equations for specific values of the physical parameters of the problem (the conductivities k_x and k_y), of the external sources (Q) and of the boundary conditions (the prescribed heat flux q over the contour Γ_q and the value of the prescribed temperature $\overline{\varphi}$ over the contour Γ_φ). In reality, the technician is not satisfied with the mere solution of the Eqs. (1), but it also requires that said solution be obtained in the most accurate, simple and economical way possible.

General Scheme of the Numerical Solution of a Problem

The numerical solution of a problem governed by differential equations with their boundary conditions is based on finding an approximate solution to the unknowns of the problem u. Said approximate solution is called $\hat{u}$ and is expressed, in general, in terms of products of known approximation functions $N_i(x)$ (generally of polynomial nature) and unknown coefficients a_i. Thus, it can be written:

$$u \cong \hat{u} = \sum_{i=1}^{N} N_i\left(x\right) a_i \tag{7}$$

The previous stage is known in numerical jargon as discretization of the solution of the continuous problem. The problem now focuses on knowing the N vectors a_i that are the unknowns of the problem. The transition from the governing differential equations to an algebraic system of equations is usually carried out through the weighted residual method (*MRP*) (Finlayson & Scriven, 1965). This procedure can be easily understood by substituting approximation (7) into the governing equations (1).

This leads to:

$$\mathrm{A}\hat{\mathrm{u}} - \mathrm{b} = \mathrm{r}_\Omega \text{ in } \Omega \tag{8}$$

$$\mathrm{B}\hat{\mathrm{u}} - \mathrm{t} = \mathrm{r}_\Gamma \text{ in } \Gamma. \tag{9}$$

The vectors r_Ω and r_Γ are called residuals of the governing equations on the domain and the boundary, respectively. The MRP is simply based on imposing that these residuals cancel out in a weighted way over Ω and Γ. The resulting MRP expression is:

$$\int_\Omega w_i^T d_\Omega + \int_\Gamma \overline{w}_i^T r_\Gamma d\Gamma = 0 \tag{10}$$

In the Eq. (10) w_i and $\overline{w}_i$ are vectors containing arbitrary weight functions defined on Ω and Γ, respectively. Substituting into (10) the expressions for r_Ω and r_Γ from (8) and (9) we obtain:

$$\int_\Omega w_i^T \left[A\hat{u} - b\right] d_\Omega + \int_\Gamma \overline{w}_i^T \left[\left[B\hat{u} - t\right] d\Gamma = 0 \tag{11}$$

Now choosing N vectors of weight functions w_i and $\overline{w}_i$, we obtain an algebraic system of N equations with N unknowns, which in stationary problems can be written in the form:

$$K a = f \tag{12}$$

In Eq. (12)K is the so-called rigidity matrix of the system, a is the vector of unknowns that contains the N vectors a_i and f is a vector that contains known terms coming from the vectors b and t in the governing equations. The solution of the algebraic system of equations (12) provides the N vectors ai. Substituting these into Eq. (7) the approximate values of the unknowns u at each point of the domain are obtained. Obviously, the expressions of the matrix K and the vector f depend on the selection of the weight functions. In the following section, the most common numerical methods are classified according to the type of weight functions chosen.

Classification of the Most Usual Numerical Methods

It is very interesting to deduce the different numerical methods used in engineering practice through the selection of the weight functions w_i and w_i. A brief, non-exhaustive classification is then made.

Finite Difference Method

The finite difference method (*FDM*) is based on choosing for w_i and $\overline{w}_i$, Dirac delta functions in a series of points selected from the analysis domain (Kasper, 2001) usually ordered according to an orthogonal mesh. This is equivalent to placing the differential equations at these points. The resulting expressions are:

$$\left[A\hat{u} - b\right]_i = 0 \; i = 1, N_\Omega$$

(13a)

$$\left[B\hat{u} - b\right]_j = 0\, j = 1, N_\Gamma$$

(13b)

where N_Ω and N_Γ are the number of placement points in the domain Ω and in the contour Γ, respectively. The derivatives that appear in operators A and B are generally evaluated by difference formulas, depending on values of the unknowns at points adjacent to each point where the equations are placed. It is highlighted that the resulting matrix K in the *MDF* is not, in general, symmetrical.

Finite Element Method (FEM)

In the finite element method (*FEM*) the analysis domain is divided into a mesh (regular or irregular) formed by simple geometric figures (triangles and quadrilaterals in 2D and tetrahedra and hexahedrons in 3D). The approximation of the unknowns (7) is now carried out inside each element, so that the parameters ai become the values of u at specific points of the element (called nodes). In the *FEM* it is usual to choose the same approximation functions *Ni* for the components of w_i and w_i (Galerkin method). This leads in most cases, after an appropriate integration by parts of some terms of A, to a symmetric expression of the matrix K. The local nature of the *FEM* approximation also means that the matrix K and the vector f of eq. (12) can be easily obtained by assembling the individual contributions of the different elements, which greatly simplifies the calculation process (Hawkes & Kasper, 2018).

Figure 3 shows schematically the analysis process of a bridge structure by the *FEM*. From the initial geometry of the bridge, the rectangular finite element mesh is obtained that discretizes the upper slab and the beams. Once the calculation process is completed, complete information is obtained on the behaviour of the bridge under the acting loads, such as, for example, the change in geometry under the loads, deformations and stresses at each point, etc. (Onate, 2013).

The "discretization" process of the bridge is conceptually similar to that used by Archimedes to calculate the perimeter of the circle by dividing it into polygons. Likewise, the rectangles in Figure 3 also play a role analogous to those used by Chinese calculators to evaluate the area of the circle in the *5th* century *AD*.

The finite element approach offers a logical step up from the analysis of bar structures using matrix calculation methods to that of continuous type structures from the perspective of a structural engineer. In fact, the first attempts to use matrix methods to solve problems with two-dimensional elasticity—that is, by breaking the continuum up into bar components—arose in the early 1940s. R. Courant (2016) first introduced the idea of a "continuous element" in 1943 to solve problems requiring planar elasticity by dividing the analytic domain into triangular "elements" and assuming a polynomial variation of the answer on each.

Figure 3. Analysis of the deformation of a bridge by the finite element method

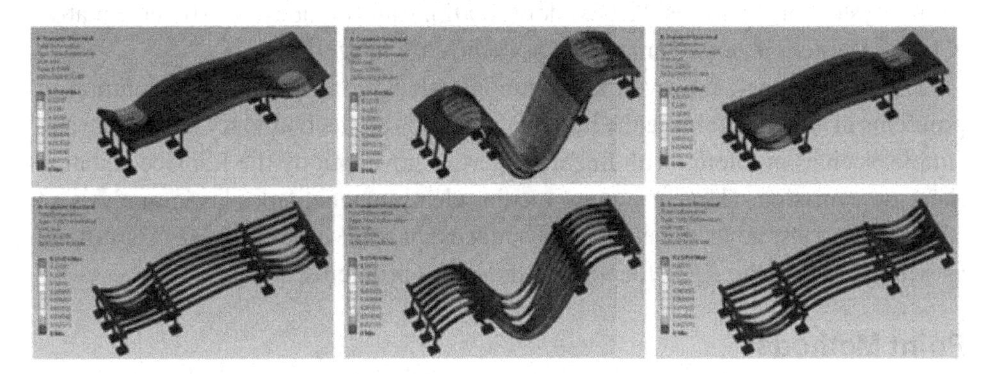

When digital computers became widely available in the 1960s, there was an incredible advancement in numerical methods that made use of matrix techniques. This led to the elimination of the limitations that came with solving large systems of equations till the present day. In this time, the finite element method quickly became the go-to solution for solving a variety of physics and engineering problems. Notably, in this specific context, structural calculation challenges—moreover, in the field of aeronautical engineering—are what are driving the early implementations of this technique (Argyris & Kelsey, 1960). R. Clough coined the phrase "finite elements" in 1960. In the process of resolving plane elasticity issues, this phrase was created (Sodhani & Reese, 2012). Since then, the *FEM's* applicability to numerous other fields has advanced remarkably. As a result, the *FEM* now occupies a unique position as a powerful technique for resolving a broad variety of complex issues in several engineering and scientific disciplines, mostly due to advancements in digital computers. These include analysing the aerodynamics of aeroplanes, investigating the historical structural integrity of structures, estimating the resistance of human bones, and assessing vein blood flow.

Boundary Element Method

In the boundary element method (*BEM*), the weight functions w_i are chosen so that they cancel over the domain Ω. The resulting integral expression is defined only on the contour as:

$$\int_{\Gamma} \overline{w}_i \left[\left[B\hat{u} - t \right] d\Gamma \right. = 0 \tag{14}$$

The *MEC* only requires, therefore, approximating the unknowns on the contours of the analysis domain, which leads to a significant decrease in the size of the resulting system of equations (Katsikadelis, 2002). Unfortunately, the stiffness matrix K of said system is not symmetrical, as is the case in *MDF*.

The *MEC* allows, however, a substantial calculation saving in the solution of problems in which obtaining the integral equations on the boundary is possible and simple, such as problems with linear properties governed by the Laplace equation (heat transmission, flow in porous media, electromagnetism, elastic analysis of solids, etc.). Despite its undeniable advantages, in some cases, the *MEC* is still not as widely used as the *FEM* for solving industrial problems.

Point Methods

We cannot conclude this brief review on numerical methods without mentioning the recent rise of methods based on discretization's using only a set of points. These methods, commonly referred to as meshless (Katsikadelis, 2002; Barazanchi et al., 2021), particle or finite point methods (*MPF*), have the advantage of not requiring the usually expensive construction of a mesh over the analysis domain; It is enough to "fill" its interior with a large number of points to which the values of the unknowns of the problem are associated. The set of points close to a specific point is called a "cloud." The variation of each unknown inside a cloud is expressed as a function of the variables at each point of the cloud using weighted least squares techniques. The final step is to impose compliance with the differential equations governing the problem on each cloud (in integral form), or directly "placing" the equations in each of the points that discretize the domain, in a similar way to the *SDM*. In both cases, the desired algebraic system of equations is reached, whose solution leads to the numerical values of the unknowns at each point (Pavel & Rachchh, 2020).

Typology of Equations in Engineering Problems

Schematically, the equations involved in the most common engineering problems can be classified into the following three large groups.

a. **Static problems:** The equations have the form of expressions (1), and in them all variables and parameters are independent of time. The system of algebraic equations after the discretization process has the matrix form expressed by Eq. (12).

Examples of engineering applications of statics problems are very numerous in the calculation of structures and mechanical systems, in heat transmission in steady state, in problems of filtration and electromagnetism in steady state, in stationary problems of fluid dynamics, etc.

a. First order dynamic problems

The general form of the governing equations in this type of problem can be expressed as:

$$A(u, \acute{u}) - b(t) = 0 \; in \; \Omega \tag{15a}$$

$$B(u, \acute{u}) - q(t) = 0 \; in \; \Gamma \tag{15b}$$

Where $\acute{u} = \frac{d}{dt}u.$ *being t the time. In this case* $u \equiv u\left(x_i, t\right)$
An example of the previous expressions is the equation of heat propagation by conduction in a transient regime. In one dimension,

$$\frac{\partial\varphi}{\partial t} + \frac{\partial}{\partial x}\left[\left(k\frac{\partial\varphi}{\partial t}\right) + Q = 0 \; in \; \Omega \tag{16a}$$

$$k\frac{\partial\varphi}{\partial x} - \bar{q} = 0 \; in \; \Gamma_q \tag{16b}$$

$$\varphi - \bar{\varphi} = 0 \; in \; \Gamma_\varphi$$

$\varphi(x, 0) = 0$ for $t = 0$

In (16a) ρ and c are the density and specific heat, respectively. The form of the resulting system of equations after the discretization process over the spatial domain Ω can be expressed as

$$Ca + Ka = f \tag{17}$$

The eq. (17) is a system of algebraic equations where the paras intervene unknown meters of the problem a and its derivatives with respect to time a. The next stage in the numerical solution of the problem is the temporal integration of Eq. (17). This can be done using different techniques for numerical solution of parabolic differential equations.

The most popular procedure to integrate Eq. (17) is to apply the finite difference method. In any case, the temporal integration procedure provides the values of the unknown parameters at time n+1 as a function of values of said parameters known at time n. Generally

$$a^{n+1} = g(a^n) \tag{18}$$

There are different techniques to obtain the vector g(an), either by solving an algebraic system of equations at each time step (implicit methods), or directly without the need to solve a system of equations (explicit methods) (Wu et al., 2014).

The engineering applications of this class of differential equations are also very numerous. Among the most characteristic problems we can mention all those of fluid dynamics in transient regime; non-stationary problems of heat transmission, filtration in porous media and electromagnetism; soil consolidation problems, etc.

b. Second order dynamic problems

The governing equations are written by

$A(u, \grave{u}, \ddot{u}) - b(t) = 0 \; in \; \Omega$

(19a)

$B(u, \acute{u}, \ddot{u}) - q(t) = 0 \; in \; \Gamma$

(19b)

A typical example is the wave equation. In one dimension,

$$\rho\frac{\partial^2\varphi}{\partial t^2}+c\frac{\partial\varphi}{\partial t}+\frac{\partial}{\partial x}\left[\left(k\frac{\partial\varphi}{\partial t}\right)+Q = 0 \; in \; \Omega\right. \tag{20}$$

$$k\frac{\partial\varphi}{\partial x}+\overline{q} = 0 \; in \; \Gamma_q$$

$$\varphi-\overline{\varphi} = 0 \; in \; \Gamma_\varphi$$

φ (x, 0) = 0 for

$$t = 0 \tag{21}$$

The temporal discretization process leads to an algebraic system of equations of the form

$$M\ddot{a} + Ca + Ka = f \tag{22}$$

The numerical solution in time by a finite difference algorithm proportional tions the values of a, a, and ä at time n +1 as a function of their values at previous times.

The practical applications of equations of this type are very common in dynamic analysis of solids and structures and in wave propagation problems (acoustics, electromagnetism, waves, etc.).

NUMERICAL METHODS AND REALITY

It is clear that the objective of numerical methods is to reproduce as closely as possible the behavior of the world through numbers. Let us remember, however, that the first step in this process is to establish a mathematical model of the reality in question. Numerical methods allow the mathematical equations of said model to be solved numerically with the help of the computer, generally expressed through equations or inequalities in partial or algebraic derivatives. The numbers that result from the calculation, expressed through computer-drawn graphics, represent the vision of reality provided by the chosen calculation process. Obviously, the numerical solution will only coincide with reality if:

a) the mathematical model incorporates all aspects of the real world.
b) the numerical method can exactly solve the equations of the mathematical model.

In practice, neither of these two conditions is met and it must be admitted that the numerical prediction will not match real-world behavior. It is then said that the numerical solution approximates reality. If we know the "real" solution of the problem being studied, we can compare it with the numerical solution and obtain the prediction error. In practice, it also happens that generally such a real solution does not exist, since every mathematical model expresses a simplified idealization of reality. Thus, in the best of cases it is possible to obtain "exact" solutions from some mathematical models that are approximations of reality. These "exact" solutions (which we called analytical solutions at the beginning of the article) can be compared with those obtained by solving the same mathematical models by numerical methods. Unfortunately, exact solutions are also practically impossible to obtain for most mathematical models that solve problems of interest. The few cases (generally academic) in which this comparison is possible serve to calibrate the numerical method. In the rest of the situations, the only feasible comparison is achieved with experimental test results obtained for specific problems, in which it is possible to make measurements. Naturally, experimental validation is useful to calibrate both the numerical method used and the underlying mathematical model.

In summary, our vision of reality will always be approximate, both due to the limitations of formalizing said reality through a mathematical model and due to the errors inherent in the application of numerical methods to said model. Our only possible reference is the empirical validations of the numerical results, using experimental values obtained in very specific laboratory or field tests. In most cases we are alone facing the set of numbers that result from the prediction of a problem whose "real" solution is unknown. It is at that moment when all the experience accumulated in the calibration and validation of the mathematical model and the numerical method chosen must be used to accept or not the numbers provided by the calculation.

Perhaps no example like that of the estimation of the number π exemplifies the limitations of numerical methods. As is known, the number π is one of the incommensurables, that is, it has an infinite number of figures and any calculation process leads to an approximate numerical solution of the value of π. Take, for example, the method used by Archimedes, based on dividing the circle into inscribed and circumscribed polygons with an increasing number of sides. By progressively refining said calculation process, that is, taking more and more polygons inscribed and circumscribed in the circumference, the error of the approximation can be reduced and the value of π limited between two numbers that are increasingly closer. In any case, the numerical solution will always be an estimate of the real value of π, although we must accept it as useful for any subsequent calculation in which its value is needed (Figure 1).

To reveal the relationships between reality, numerical methods, mathematical models and computer science, we would have to answer many questions: What makes a numerical solution truthful? Why should I believe numerical values obtained through the computer? What makes a numerical solution useful? What makes it good or bad? What makes her beautiful or ugly? What has been the influence of numerical methods on theories of mathematical knowledge and existence, on mathematical intuition, on mathematical education? What relationship exists between the pen-mathematical knowledge, numerical methods and the potential capabilities of the computer and the human mind? How does the numerical solution of a problem contribute to changing our idea of reality, of knowledge, of time?

Once the previous questions have been answered, we would be well on our way to creating a philosophy of numerical methods (or calculus). Thus, just as classical philosophy has dealt with the true, the good and the beautiful, so too the philosophy of numerical methods could deal with the veracity of calculations, with the goodness and beauty in the numerical solution of problems of the universe. Only through deep reflection on these ideas can we shed light on a society increasingly divided between those who believe that computing and its subsequent activities are an irremediable evil, which degrades the spirit and corrupts intelligence, and those who, in The opposite pole, believe that the numbers provided by the computer, generated through mathematical models of reality and numerical methods, help us better understand the world around us and are one more ingredient to achieve social justice and peace. world.

APPLICATIONS

In this section, various applications of the finite element method to engineering problems are presented. The first application refers to the analysis of historical structures. Figure 4.1 shows the discretization into hexahedral elements of the five domes of St. Mark's Basilica in Venice. The non-linear structural analysis consisted of calculating the safety coefficient against breakage of the structure obtained by progressively increasing the weight until the domes broke. Figure 4.2 shows the different levels of deterioration in the structure at the time of ruin. The intensity of the deterioration is characterized by the range of colors.

Figure 4. Discretization of the domes of St. Mark's Basilica with hexahedral finite elements

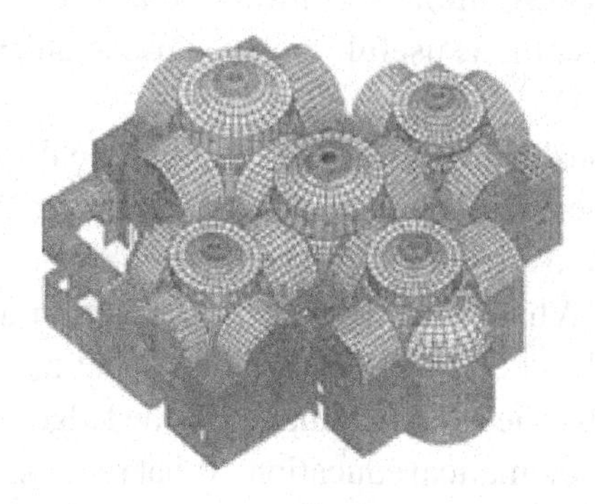

Deformed structure under its own weight.

Figure 5. St Mark's Basilica—Levels of deterioration in the structure for a load equivalent to 7.11 times its own weight

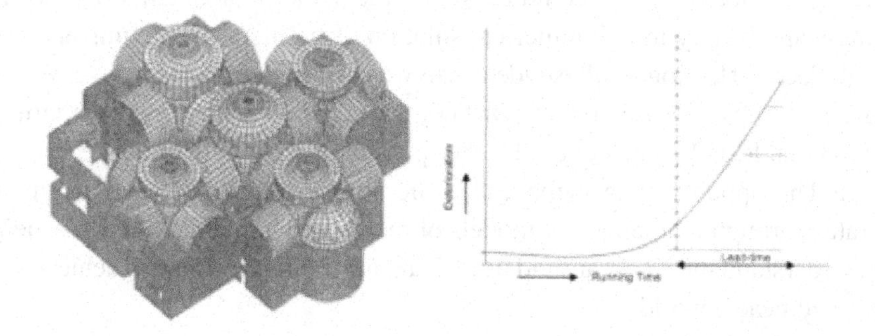

Figure 6. Structural analysis of the Barcelona Cathedral

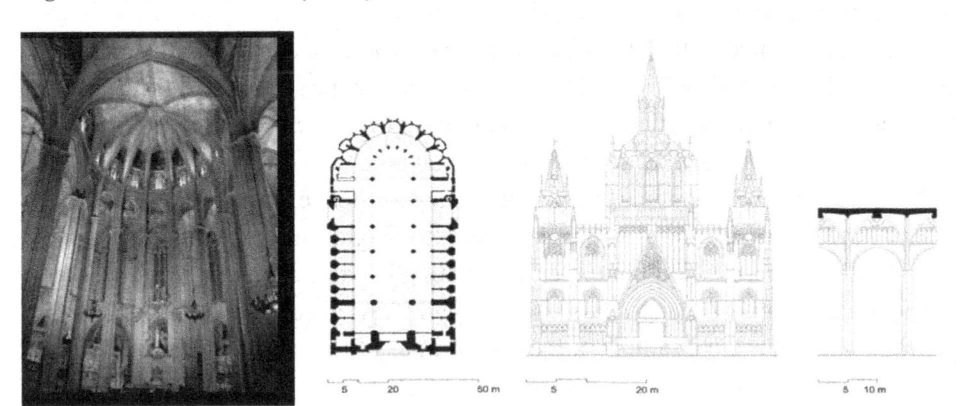

An example of a similar study carried out on the Barcelona Cathedral is shown in Figure 5. For more details on analysis by the FEM of historical structures (Barbat et al., 1997; Shawkat et al., 2022).

The following example shows different results of the aerodynamic study carried out on the new building of the Gran Telescopio de Canarias, located on the island of La Palma. The outline of the building that houses the telescope (Figure 6) is discretized into finite elements. After this, the volume of air inside and outside the building is discretized into tetrahedra and the Navier-Stokes equations that govern the air flow inside a control domain that includes the building and its environment are solved. As results, the fields of velocities, pressures and temperatures at all points of the domain are obtained. These results are used to design the building's ventilation system, through the study of the trajectory of air particles, to evaluate the forces on the different instruments, to obtain the temperature in the vicinity of the telescope mirrors and to estimate the possible disturbance of the image due to the effect of turbulence in the fluid (Takahashi, 2015). The following example presents the hydrodynamic and aerodynamic analysis of a sailboat. In the study by the FEM, the air flow around the sails and the hull and the hydrodynamics of the submerged part of the latter are analyzed. Different results of this study are shown in Figure 7.

The following example is an analysis of the manufacturing process of a car crankshaft by casting. The study involves simulating in detail the process of filling the mold with molten metal and then analyzing the solidification and cooling process of the metal, from the temperature of 1200° of the molten steel to room temperature. As a result of the study, the distribution of deformations and stresses in the piece during the solidification and cooling process is also obtained. Figure 8 shows the temperature maps in the crankshaft at different times of the cooling process. For more information about this problem and other applications of FEM to metal forming processes, consult the references.

The last example is the study of the deformation of a car in a collision with a rigid wall. The analysis was carried out by discretizing the structure of the car by the FEM and solving the equations of the dynamics of the structure, taking into account the effects of the contact between the vehicle and the wall during the crash. Figure 9 shows the finite element mesh used and a view of the deformation of the vehicle after 24 milliseconds after the impact began. It is highlighted that the results have been obtained on a dual PC computer, calculating in parallel on the two processors, through a partition of the structure into two domains as shown in the figure. The previous examples are a sample of the possibilities of the FEM in its application to the industrial world.

Figure 7. Aerodynamic analysis by the FEM building. Discretization of the building and the surrounding terrain and different results of the distribution of pressures and velocities and the trajectories of air particles

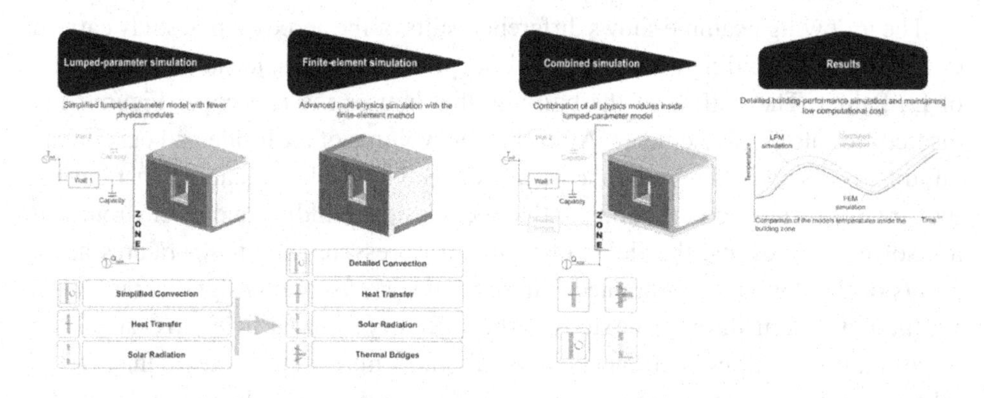

Figure 8. Analysis by the FEM of the aerodynamics and hydrodynamics of a competition sailboat

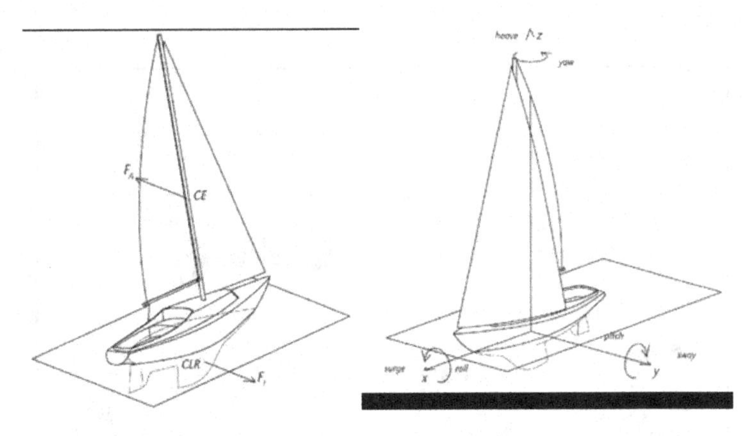

Figure 9. FEM analysis of the solidification and cooling process of a crankshaft manufactured by casting

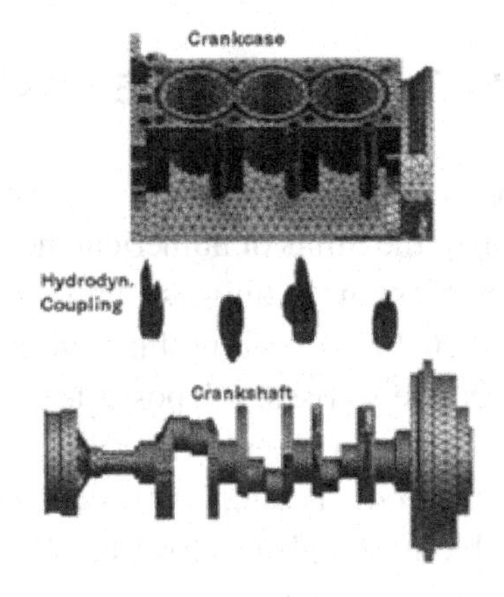

Figure 10. Finite element mesh, partition of the structure into two domains for parallel calculation and deformation of the vehicle after 24 milliseconds from the start of the impact

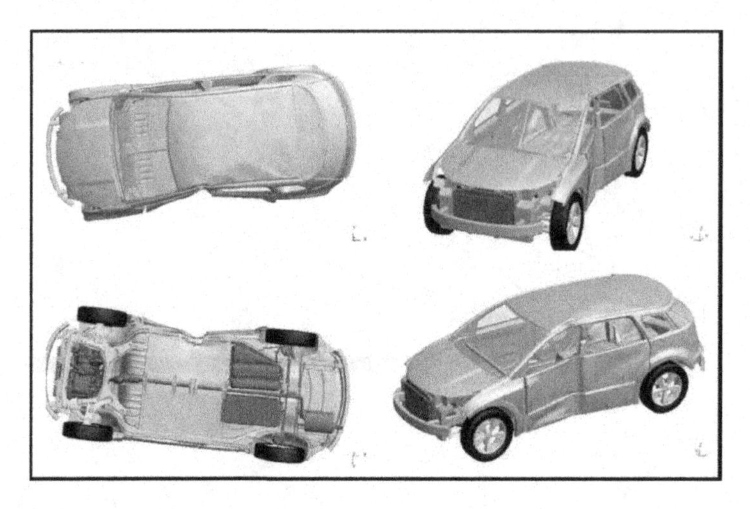

LIMITS AND PERSPECTIVES OF NUMERICAL METHODS

Is it possible to describe all aspects of the world with the help of mathematics and numerical methods? In reality, the limits of numerical methods, that is, the limits of being able to express any problem in the universe numerically, are closely linked to the possibility of posing (formalizing) said problem in mathematical form. In this situation, the question immediately arises: Is it possible to mathematize everything? Is there anything in the world that can never be described in mathematical language? As P. J. Davies and R. Hersh (Codina et al., 2000) state, there does not seem to be anything non-mathematizable in the physical world. We are confident, therefore, that physics is capable of encompassing any physical phenomenon and that it is capable of doing so through a mathematical formalism. The limit of what is clearly mathematizable seems, therefore, to coincide with the limit of the physical world. What another world is there? If our convictions are those of a pure mechanistic materialist, we will probably answer none.

However, it is evident that there are such things as emotions, beliefs, attitudes, dreams, intentions, jealousy, envy, longings, regrets, feelings such as anger and compassion and many others. These things, which make up the inner world of the human mind, and, even more, all those that encompass the "inner life" of society, of civilization itself, for example, literature, music, politics or the tides and currents of history, can they be mathematized?. The answer is neither simple nor banal. Throughout history we find defenders of the belief that everything is mathematizable, and therefore numeralizable, and the opposite.

Taking sides, we could affirm that all facets of the world that can be framed by concrete statements can be described with mathematical models. Naturally, if said statement is essentially humanistic, the difficulty of establishing the model will be much greater. Take, for example, the feeling of well-being. In this case, an experimentalist could claim that well-being is a mere function of hormone and blood sugar levels, once again suggesting the possibility of finding a mathematical/numerical model for well-being. An old conflict between scientists and humanists comes from the fact that the latter feel that there should be a portion of the world immune to mathematization, while scientists feel the opposite, that is, every aspect of the world admits a mathematical and therefore numerical description. Once again, the old aspiration of Pythagoras and Plato reappears, gaining followers as new successes occur in the mathematization of the world.

Nowadays, the bases of natural sciences, physics, astrophysics, chemistry and also all engineering are deeply mathematical. Mathematical and numerical models are, in all these disciplines, tools of daily use, both for the explanation of phenomena in the universe and for the design of products and processes. Furthermore, for a new theory to be considered scientific, and therefore accepted, it is almost a necessary condition that it be expressible in mathematical language, and it is almost certain that, if the available mathematics is inadequate to describe certain observed phenomena, it will be possible to devise and develop the appropriate ones.

The life sciences, biology and medicine, are becoming increasingly mathematical/numerical in nature. The mechanisms that control physiological processes, genetics, morphology, population dynamics, epidemiology and ecology have been provided with mathematical and numerical models. It is not possible to understand economic theory without a solid mathematical and numerical training. Theories of competition, trade cycles and balances require mathematics and quantification techniques of a deeper type. Determining a trade or military policy may require decision theories, game theories, and optimization strategies.

Regardless of the problem being solved, it is important to remember that the ultimate goal of numerical methods is to provide understanding, not numbers. In short, numerical methods must be closely linked to both the source of the problem and the intended use of the numerical results; The application of these methods is not, therefore, a stage to be considered in isolation from reality (Jiang et al., 2014). "Multiphysics" is a word that properly sums up the future promises of numerical techniques in multiple domains. Problem-solving will no longer be restricted to a single physical medium but will instead take into account all of the linkages that characterise reality's complexity. As a result, while designing a part for a vehicle, like an automobile or aeroplane, great thought is given to the manufacturing process as well as the component's intended use over the duration of its life. The topography, water, and air surrounding a structure are all factors that are taken into account when studying it in civil engineering. Similar cases may be found in almost every scientific subject, including bio-engineering, economics, demography, modelling human and collective behaviour, and the domains of naval and aeronautical engineering. It is imperative to take into account the non-deterministic character of all data in order to precisely evaluate the likelihood that recently created mathematical models will behave as expected.

Close collaboration amongst all members of the triangle—which is predicated on a thorough comprehension of the mathematics and physical underpinnings of each problem—is necessary to find effective solutions to the megaproblems of the twenty-first century. It is essential that computer science and numerical methodologies work together. A greater emphasis on optimising the distribution of material and human resources necessary to successfully handle the growing scope of issues that need to be resolved should also be a sign of this partnership. It should also place a high priority on putting creative training programmes into place so that the next generation of workers is prepared to use mathematics and numbers to solve interdisciplinary problems.

REFERENCES

Argyris, J. H., & Kelsey, S. (1960). *Energy theorems and structural analysis* (Vol. 60). Butterworths. 10.1007/978-1-4899-5850-1

Bidwell, J. K. (1994). Archimedes and Pi-Revisited. *School Science and Mathematics*, 94(3), 127–129. 10.1111/j.1949-8594.1994.tb15638.x

Clough, R. W. (1960). *The finite element in plane stress analysis. Proc. 2^< nd> ASCE Confer.* On Electric Computation.

Codina, R., Morton, C., Oñate, E., & Soto, O. (2000). Numerical aerodynamic analysis of large buildings using a finite element model with application to a telescope building. *International Journal of Numerical Methods for Heat & Fluid Flow*, 10(6), 616–633. 10.1108/09615530010347196

Finlayson, B. A., & Scriven, L. E. (1965). The method of weighted residuals and its relation to certain variational principles for the analysis of transport processes. *Chemical Engineering Science*, 20(5), 395–404. 10.1016/0009-2509(65)80052-5

Guicciardini, N., Kjeldsen, T. H., & Rowe, D. E. (2006). Mathematics in the Physical Sciences, 1650-2000. *Oberwolfach Reports*, 2(4), 3175–3246. 10.4171/owr/2005/56

Huang, J., Yang, R., Ge, H., & Tan, J. (2021). An effective determination of the minimum circumscribed circle and maximum inscribed circle using the subzone division approach. *Measurement Science & Technology*, 32(7), 075014. 10.1088/1361-6501/abf803

Imanova, G. (2022). History of Physics. *Journal of Physics & Optics Sciences.* doi.org/ 10.47363/JPSOS/2022

Kasper, E. (2001). The Finite-Difference Method (FDM). *Advances in Imaging and Electron Physics*, 116, 115–191. 10.1016/S1076-5670(01)80068-9

Kasper, E. (2018). The Finite-Difference Method (FDM). In *Advances in Imaging and Electron Physics*. Elsevier.

Katsikadelis, J. (2002). *Boundary Elements. Theory and Applications*. Elsevier.

Oñate, E. (2013). *Structural analysis with the finite element method. Linear statics: volume 2: beams, plates and shells.* Springer Science & Business Media.

Patel, V. G., & Rachchh, N. V. (2020). Meshless method–review on recent developments. *Materials Today: Proceedings*, 26, 1598–1603. 10.1016/j.matpr.2020.02.328

Shawkat, S. A., Ismail, R. N., & Abdulqader, I. R. (2022, November). Implementation a hybrid ADHOC sensor system. In *AIP Conference Proceedings* (*Vol. 2394*, No. 1). AIP Publishing. 10.1063/5.0121146

Tuama, B. A., Shawkat, S. A., & Askar, N. A. (2022, November). Recognition and classification of facial expressions using artificial neural networks. In *Proceedings of Third Doctoral Symposium on Computational Intelligence: DoSCI 2022* (pp. 229-246). Singapore: Springer Nature Singapore.

Wu, N. J., Chen, B. S., & Tsay, T. K. (2014). A review on the modified finite point method. *Mathematical Problems in Engineering*.

Chapter 4
Handling Irregular and Complex Geometries

C. B. Sivaparthipan
Tagore Institute of Engineering and Technology, India

WaleedKhalid Al-Azzawi
Al-Farahidi University, Iraq

ABSTRACT

In the analysis using the finite element method (FEM) of structural components under multiple loading conditions, these are studied individually to obtain the individual structural responses of each of them. In a standard h-adaptive analysis process, different meshes would be generated with each applied load. Therefore, an extensive analysis would be necessary to obtain the stiffness matrix for each case. A solution to this approach is based on receiving a mesh that can be used for all cases and that, in turn, allows the user to achieve the desired error. This work defines a practical method that allows obtaining the optimal mesh in linear static refinement problems for multiple load cases about the size relationships of the elements and the expressions that relate the estimated errors between successive meshes. The suggested method is numerically validated through two-dimensional examples.

INTRODUCTION

In the analysis of problems in structural engineering, it is expected to find cases in which each of the components of a system is subjected to multiple loading conditions. Analyzing structural components under various loading conditions enables engineers to comprehensively assess performance, identify critical scenarios, understand load interactions, validate design assumptions, optimize design, and facilitate

DOI: 10.4018/979-8-3693-3964-0.ch004

risk assessment & mitigation. This provides a holistic understanding of structural behavior and allows for safer, more efficient, and resilient structures. When the study is done by the finite element method (FEM), these loads are separated in different cases to obtain the individual structural response of each of them. FEM helps analyze structures under various loading conditions by using the principles of superposition, incremental loading, coupled analysis, time integration, and load combinations. By employing these techniques, FEM can handle complex and diverse loading environments and assess the overall structural performance and safety. A linear analysis would imply a slight increase in computational cost because the same finite element mesh can be used for all load cases. However, in a standard h-adaptive analysis process, different meshes would be generated with each applied load. Generating a single mesh that can be used for all load cases in h-adaptive analysis is challenging due to load case compatibility, solution sensitivity, interactions between load cases, computational efficiency, and mesh quality and stability. Achieving this often requires advanced techniques, validation, and iterative refinement. The computational cost difference between linear analysis and standard h-adaptive analysis in structural engineering arises from differences in mesh refinement strategies and computational requirements. H-adaptive analysis results in a finer mesh in regions of interest and coarser mesh elsewhere, improving accuracy and efficiency. However, it involves iterative processes for mesh refinement and solution re-computation, which add computational overhead. Linear analysis requires less effort in mesh generation and refinement, making it more resource-efficient and suitable for applications with limited computational resources. Therefore, an extensive analysis would be necessary to obtain the stiffness matrix for each case. A solution to this approach is based on receiving a mesh that can be used for all cases and that, in turn, allows the user to achieve the desired error. A versatile mesh generates meshes for various loads across different applications. It should adapt to different structural components and configurations, support a variety of element types, incorporate flexible mesh refinement strategies, and consider the specific characteristics and demands of different load cases. It should also offer mesh generation automation, scalability, efficiency, and interoperability. Engineers and analysts can effectively address the complexities of generating meshes for various loads across different applications by employing a versatile mesh that offers adaptability, scalability, and interoperability.

The issue of mesh adaptivity in finite element method solutions has been widely discussed in recent years, and procedures for generating optimal meshes for a load case in static problems have been developed (Erhunmwun, I. D., et al., 2017), (Carson, H. A., et al., 2017). Today, they are used as practical tools. Mesh adaptivity is crucial in finite element method solutions for structural analysis because it improves solution accuracy, optimizes computational resources, conforms closely to complex geometries, resolves singularities, enables dynamic adaptation to changing conditions, refines the mesh iteratively, and facilitates error control and convergence monitoring. Generating optimal meshes for load cases in static problems involves considering specific criteria to ensure accurate and efficient solutions. The critical criteria for generating optimal meshes are solution accuracy, element aspect ratio, boundary layer resolution, mesh gradation, solution convergence, computational efficiency, and mesh adaptivity. By considering these criteria, engineers can generate optimal meshes for load cases in static problems, ensuring accurate and efficient solutions in structural analysis and design. The basic idea of these procedures consists of generating optimal meshes by iteratively applying criteria for uniform distribution of the discretization error. The iterative process for generating optimal meshes involves the following steps: initial mesh generation, error estimation, error distribution analysis, mesh refinement, solution re-computation, error assessment, convergence check, and optimal mesh selection. The process refines the mesh iteratively based on error distribution analysis and solution re-computation until a satisfactory solution is obtained. This approach enables the generation of optimal meshes that achieve a uniform distribution of discretization error and accurately capture the solution behavior with minimal computational resources (Pacheco, J. P. F. R., et al., 2020) or techniques that seek to minimize the number of degrees of freedom in the new mesh. However, although the adaptivity criteria are well-defined when a static load case is considered, the same is different when several load cases are considered (Li, J., et al., 2020).

Despite the significant advances in developing mathematical theory and algorithms for the Finite Element Method, the discretization of a given problem depends, in most cases, on the analyst's common sense and the experiences acquired in solving other problems. The discretization must be redone if the results are judged to be wrong. Therefore, it is reasonable to assume that if the analyst's intuition failed when designing the model, the same could happen when judging the validity of the results1.

Due to these uncertainties, the possibility of automatically improving the quality of a numerical solution has become a question of great interest in Computational Mechanics. The approximate solution, obtained through the Finite Element Method, can be improved by implementing adaptive strategies, which consist of automatically modifying the model in domain regions where the accuracy is unsatisfactory. Adaptive strategies in the Finite Element Method enhance the accuracy of the approximate solution by refining the mesh locally, reducing the discretization error, adapting the mesh to changing solution characteristics, resolving singularities, optimizing mesh quality, and efficiently utilizing computational resources. Thus, in this work, the adaptive versions h, p, and hp of the Finite Element Method are presented in the context of the elastic bending problem of Reissner-Mindlin plates.

ADAPTIVE STRATEGIES

Adaptive refinement techniques for finite element meshes are generally based on changing the location of nodes without changing the mesh topology (refinement r), on refining the mesh by increasing the number of elements (refinement h), on improving the polynomial order of the elements (p refinement) or through combinations of these; mainly from h and p refining (hp refining). Adaptive refinement techniques for finite element meshes involve generating a coarse mesh, computing the solution, estimating errors, and refining the mesh based on error indicators. The mesh is refined locally in regions of interest to improve solution accuracy. This process is repeated until a desired level of accuracy is achieved, resulting in more accurate and efficient finite element solutions.

In adaptive methods, the ability to locally evaluate the discretization error is implicit, providing a measure of the quality of the approximate solution. Adaptive methods for improving solution accuracy can be achieved using "h-refinement," "p-refinement," and "hp-refinement" techniques. H-refinement subdivides existing elements, p-refinement increases the degree of approximation within individual components, and hp-refinement combines both methods. Each method has advantages and disadvantages, and the choice between them depends on the specific requirements of the problem being solved. The refining techniques mentioned above can improve the solution obtained efficiently using this information.

Version h: In mesh refinement h, the polynomial order of the element interpolation functions remains constant while their size is modified depending on a locally calculated error estimate. The polynomial order of element interpolation functions during mesh refinement is influenced by several factors: element type, refinement strategy, mesh quality, solution smoothness, boundary conditions, computational cost, and numerical stability. Engineers and researchers can consider these factors to make informed decisions about the polynomial order to achieve accurate and efficient numerical simulations.

The adaptive version is suitable for cases where the solution is not smooth throughout the domain, such as interfaces between different materials, boundary layers, the boundary of a plastic front, etc. Adaptive mesh refinement can result in non-smooth solutions due to discontinuities, interpolation errors, numerical artifacts, mesh transition zones, solution sensitivity, physical phenomena, and convergence criteria. Addressing these issues requires careful criteria selection, improved interpolation methods, and post-processing techniques.

This work describes a practical procedure that allows obtaining the optimal mesh in linear static refinement problems for multiple load cases based on the size relationships of the elements and the expressions that relate the estimated errors between successive meshes. A systematic approach obtains optimal meshes for linear static refinement problems with multiple load cases. An initial coarse mesh is generated, and error estimation techniques are applied to identify regions where the solution is poorly represented. Based on these indicators, the mesh is refined locally in the areas of interest, increasing mesh density or adding more elements to improve solution accuracy. The refinement process is performed iteratively until a desired level of accuracy is achieved for all load cases. Inadequate mesh refinement in problems with multiple load cases can compromise the accuracy of analysis results, increase computational costs, and hinder the efficiency of design optimization processes. Proper attention to mesh refinement is essential to ensure accurate and efficient structural analyses in complex engineering problems. The best mesh size for different load scenarios is determined by considering the desired solution accuracy, sensitivity analysis, error estimation techniques, load case characteristics, critical features and phenomena, computational efficiency, and convergence and stability of the numerical solution. This helps engineers achieve accurate and efficient results in structural analysis and design.

Before the procedure is described, a review of the concepts of discretization error estimation, the effectiveness of the error estimator, and the definition of the h-adaptive process are presented. These concepts are fundamental to the steps before applying the developed procedure. Once the process of obtaining the optimal mesh has been explained, the results of a selected example are presented to show the numerical validation of the developed formulation.

ESTIMATIONS OF THE DISCRETIZATION ERROR

Some procedures allow an error estimate to determine the quality of the finite element solution. One of these is the Zienkiewicz-Zhu error estimator (Mahomed, N., et al., 1998) based on the improvement of the FEM solution. This estimator is based on using, within the expression of the error in the energy standard, an improved stress field obtained from the FEM stress field. The Zienkiewicz-Zhu estimator estimates the energy norm of the error in finite element simulations. It involves a recovery operator, gradient recovery, and element-wise error estimation. The ZZ estimator is computationally efficient and applicable in linear and nonlinear analysis settings. However, its accuracy depends on various factors and might need to be revised with highly nonlinear or discontinuous solutions. From the Zienkiewicz-Zhu estimator, the energy norm of the estimated error can be obtained that $||e_{es}||$, as

$$||e_{es}||^2 = \int_{\Omega} \left(\sigma^* - \sigma_{ef}\right)^T D^{-1} \left(\sigma^* - \sigma_{ef}\right) d\Omega \tag{1}$$

Where,

σ^* *and* σ_{ef} represent the enhanced and finite element stress fields. The matrix D corresponds to the elastic constants. The relative estimated error, η_{es}, is obtained from the relationship

$$\eta_{es} = \frac{||e_{es}||}{||u_{es}||} \times 100 \tag{2}$$

where

$||u_{es}||^2$ is an estimate of the strain energy that can be obtained by taking advantage of the properties of the energy standard

$$||u_{es}||^2 \approx ||u_{ef}||^2 + ||u_{es}||^2 \tag{3}$$

where $||u_{ef}||^2$ is the strain energy of the FEM solution.

At the element level, the global relative estimated error can be defined as

$$\eta_{es}^{(e)} = \frac{||e^{(e)}{}_{es}||}{||u_{es}||}$$

$$\times 100 \tag{4}$$

Where (e) the superscript (e) indicates that the corresponding magnitude has been calculated in the element.

To study the reliability of the error estimators, the effectiveness index, T, is defined as the relationship between the values of the estimated error and the exact error.

$$\theta = \frac{||e_{es}||}{||u_{ex}||} \tag{5}$$

The error estimator is said to be asymptotically exact if the effectiveness index tends to unity when the size of the elements tends to zero. However, the effectiveness index tends to have a stable value that is different from unity. In that case, it is possible to define a correction factor for the estimator that improves the quality of the estimate. A correction factor improves estimation quality by compensating for model assumptions, adapting to non-ideal conditions, calibrating against experimental data, balancing computational efficiency and accuracy, handling uncertainty, and improving convergence properties. It allows for greater flexibility, robustness, and accuracy in estimation methods, ultimately improving estimation quality and more reliable results. In cases where the exact solution is not known, it is replaced by an approximate solution obtained from a highly refined mesh.

CONVERGENCES IN THE H-ADAPTIVE PROCESS

On the topic of mesh adaptivity in static problems with a load case, the criteria for defining the optimal mesh are widely studied. From them, procedures have been developed based on generating optimal meshes by iteratively applying criteria of uniform distribution of the discretization error or using techniques that seek to minimize the degrees of freedom in the new mesh. Reducing the degrees of freedom in the new mesh is crucial to achieving efficient and accurate simulations in h-adaptive process convergence. Strategies to address this include estimating the error, selectively refining regions where error indicators exceed a threshold, implementing a hierarchical mesh refinement, setting a threshold for the error indicators, developing error-based refinement criteria, considering solution-based metrics, and optimizing the mesh after refinement. These strategies help optimize computational resources, reduce memory requirements, and enhance the efficiency of numerical simulations. However, even though the above defines how well-established the refinement is for problems with one load case, the same only occurs when multiple load cases are considered. When considering a sequence of meshes obtained from a uniform refinement h, the exact error in the energy standard for each load case I in $2D$ problems is bounded by

$$\|e_{ex\,(i)}\| \leq C_{(i)}\left(h\right)^{min(p.\mu)} \approx C_{(i)}\left(N\right)^{1/2\; min(p.\mu)} \tag{6}$$

Where $\|e_{es\,(i)}\|$ is the exact error for the case considered, N is the number of degrees of freedom, and h is the element's size, denoted as p, determined by the degree of the polynomial of the interpolation function used. $C_{(i)}$ is a positive constant, whereas μ represents the intensity of the singularities. The asymptotic speed of convergence, denoted as the exponent of N, is contingent upon the finite element mesh and the smoothness of the solution. A unified mesh offers many benefits over generating separate meshes for each load condition. It ensures consistency and compatibility across different load cases, optimizes resource utilization, improves solution accuracy, simplifies model setup and management, enhances mesh adaptivity, and facilitates parametric studies and design optimization. A unified mesh is a practical structural analysis and design approach in engineering applications.

A mesh is deemed nearly optimum for a given scenario when the sequence of meshes is created to ensure the absolute error is distributed almost evenly throughout the elements. Distributing the absolute error evenly throughout the elements provides a balanced and accurate representation of the error across the entire domain. It prevents certain regions from dominating the error estimation process while neglecting others, leading to a fair and representative evaluation of solution accuracy. A uniform error distribution also facilitates convergence of the numerical solution process, enhances mesh adaptivity, and mitigates the effects of solution variations or uncertainties. Ultimately, this leads to improved solution quality and reliability. Additionally, the exact error is bounded by the following sequence.

$$\|e_{ex\,(i)}\| \leq C_{(i)}\left(N\right)^{1/2p} \tag{7}$$

Since the constant $C(i)$ is independent of h, N, and p, an approximate relationship can be established between the global errors for a load case and the corresponding degrees of freedom or element size for two successive meshes of the uniform refinement

$$\frac{\|e_{ex\,(i)}\|_p}{\|e_{ex\,(i)}\|_n} \approx \left[\frac{h_p}{h_{n(i)}}\right]^c \tag{8}$$

Where

$\|e_{ex\,(i)}\|_p$ and $\|e_{ex\,(i)}\|_n$ being the global errors in the previous and new meshes, respectively, hp is the size of the element in the previous mesh, $h_{n(i)}$ the new size of the component, and $c = min(p, \mu)$. In developing refinement strategies, it is assumed that the value of c is always equal to p as long as there are no singularities. If they occur, it is approximately true if the refinement is carried out adaptively.

Assuming that the law of convergence is also valid at the element level, we have

$$\frac{\|e^{(e)}_{ex(i)}\|_p}{\|e^{(e)_p}_{ex(i)}\|_n} \approx \left(r^{(e)}_{(i)}\right)^c \tag{9}$$

where $\|e^{(e)}_{ex(i)}\|_p$ is the exact error for an element of the previous mesh, $\|e^{(e)_p}_{ex(i)}\|_n$ the total error of all new elements contained in an element of the previous mesh and between the previous size of the element and the new size $r^{(e)}_{(i)}$, defined by

$$r^{(e)}_{(i)} = \frac{h^{(e)}_p}{h^{(e)}_{p\,(i)}} \tag{10}$$

The total error of all the innovative elements confined in an aspect of the previous mesh, $\left\| e^{(e)p}_{ex(i)} \right\|_p$, is related to the errors of each of the new elements by the expression

$$\left\| e^{(e)_p}_{ex(i)} \right\|_n^2 = \sum_{e=1}^{N^{(e)_p}_{n(i)}} \left\| e^{(e)}_{ex(i)} \right\|_n^2 \tag{11}$$

where $\left\| e^{(e)}_{ex(i)} \right\|_n$ is the error in an element of the new mesh contained in a component of the previous mesh, and $N^{(e)_p}_{n(i)}$ is the number of new elements contained within a component of the mesh. previous.

Rearranging the terms of equation (9), the convergence relationship at the element level is given by

$$\left\| e^{(e)_p}_{ex(i)} \right\|_n$$

$$=$$

$$\left\| e^{(e)}_{ex(i)} \right\|_p$$

$$\cdot$$

$$\left(r^{(e)}_{(i)} \right)^{-c} \tag{12}$$

Since the exact errors are not known, the estimated errors will be used. It will be assumed that the previous relations on global convergence and at the element level also hold for the estimated errors.

When examining a uniform refinement at the element level, the new magnitudes of the elements within an element of the preceding mesh are taken into interpretation, $h^{(e)}_{p\,(i)}$, is defined as

$$\left(h^{(e)}_{p\,(i)} \right)^2 \approx \left(h^{(e)_p}_{p\,(i)} \right)^2 . N^{(e)_p}_{p\,(i)} \tag{13}$$

where $h^{(e)_p}_{p\,(i)}$ is the new size of the elements that are contained in a previous element, and $N^{(e)_p}_{p\,(i)}$ is the number of new elements contained in an element of the previous mesh.

The number of new elements with size $h^{(e)_p}_{p\,(i)}$ that are contained in a previous component for a two-dimensional problem can be expressed as

$$N^{(e)}_{p\,(i)} \approx$$

$$\left(r^{(e)}_{(i)}\right)^2 \tag{14}$$

The total number of elements in the new mesh is the sum of new elements in all aspects of the previous mesh

$$N_{n(i)} = \sum_{e=1}^{N_p} N^{(e)}_{p\,(i)} \approx \sum_{e=1}^{N_p} \left(r^{(e)}_{(i)}\right)^2 \tag{15}$$

where N_p is the number of aspects of the previous mesh.

Suppose we assume that the size-element-number relationship is inversely proportional on a global scale and for uniform refinement (squared for two-dimensional problems). In that case, we can also express the total number of elements in the new mesh as a function of the elements in the old mesh.

$$\frac{N_{n(i)}}{N_p} \approx \left[\frac{h_p}{h_{n(i)}}\right]^2 \tag{16}$$

And substituting the size relationship of equation (8), obtained based on the law of convergence at a global level, we have

$$N_{n(i)} \approx N_p \left[\frac{\|e_{es\,(i)}\|_p}{\|e_{(i)}\|_d}\right]^{2/c} \tag{17}$$

The expression for the estimated absolute error in the new mesh can be derived by considering the errors associated with each of the new elements, as given by

$$\|e_{es\,(i)}\|_n^2 = \sum_{e=1}^{N_n} \|e^{(e)}_{es(i)}\|_n^2 \tag{18}$$

or using equation (11), it can also be expressed as

$$\|e_{es\,(i)}\|_n^2 = \sum_{e=1}^{N_p} \|e_{es(i)}^{(e)_p}\|_n^2 \tag{19}$$

By replacing the value $\|e_{es(i)}^{(e)_p}\|_n$ from the law of convergence at the element level, equation (12), the estimated absolute error predicted for the new mesh as a function of the size relationship is obtained.

$$\|e_{es\,(i)}\|_n^2 = \sum_{e=1}^{N_p} \left(\|e_{es(i)}^{(e)}\|_p^2 \cdot \left(r_{(i)}^{(e)}\right)^{-2c} \right) \tag{20}$$

OPTIMAL MESH CRITERIA

A FEM is optimal if it is obtained as a result of applying specific pre-established criteria so that the previously determined error values are achieved in its solution. Starting from different criteria, different strategies or procedures can be developed to obtain an optimal mesh, resulting in a mesh that is optimal under specific criteria that may not be optimal for others.

(1) Optimal mesh for a load case

The criterion traditionally used in elasticity problems for an analysis case (a load case in the static problem) is to distribute the error energy in the mesh uniformly. This has been done differently (considering the previous mesh or the new one, using relative or absolute errors). Recent research has demonstrated that distributing the absolute energy of the error in the new mesh yields a result similar to minimizing the number of elements in the new mesh (Coorevits, P., et al., 2004), (Oñate, E., et al., 1993).

The estimated inaccuracy for each element in the new mesh will be defined based on a uniform distribution of energy as below,

$$\|e_{es(i)}^{(e)}\|_n^2 = \frac{\|e_{(i)}\|_d^2}{N_{n(i)}} \tag{21}$$

and the value of the error energy in the element of the new mesh being constant, equation (11) can be written below

$$\|e_{es(i)}^{(e)_p}\|_n^2 = N_{p\,(i)}^{(e)_p} \times \|e_{es(i)}^{(e)}\|_n^2 \tag{22}$$

Substituting equation (21) into equation (22), we have

$$\left\| e_{es(i)}^{(e)_p} \right\|_n^2 = N_{p\,(i)}^{(e)_p} \times \frac{\left\| e_{(i)} \right\|_d^2}{N_{n(i)}} \tag{23}$$

The number of new elements in the mesh and in a previous element, $N_{n(i)}$, $N_{p\,(i)}^{(e)_p}$ respectively, have been defined in equations (14) and (17) and making the corresponding substitution in the equation (23) we arrive at the relationship

$$\left\| e_{es(i)}^{(e)_p} \right\|_n^2 = \left(r_{(i)}^{(e)} \right)^2 \times \frac{\left\| e_{(i)} \right\|_d^2}{N_p \left[\frac{\left\| e_{es\,(i)} \right\|_p}{\left\| e_{(i)} \right\|_d} \right]^{2/c}} \tag{24}$$

Finally, substituting the value $\left\| e_{es(i)}^{(e)_p} \right\|_n$ into the law of convergence at the element level, equation (12), and solving for the size relationship, we obtain

$$r_{(i)}^{(e)} = \left[\frac{\left\| e_{es(i)}^{(e)} \right\|_p}{\left\| e_{(i)} \right\|_d} \right]^{1/c} \times \left(N_p \right)^{1/2(c+1)} \times \left[\frac{\left\| e_{es(i)} \right\|_p}{\left\| e_{es(i)}^{(e)} \right\|_p} \right]^{1/c\,(c+1)} \tag{25}$$

Once the value of the size ratio for all elements is known, the new sizes are found directly using equation (10).

(2) Optimal mesh for various load cases

Equation (25) defines the optimal mesh for each load case, but naturally, the optimal mesh for one case does not have to be optimal for the others; this is illustrated in Figure 1, where the behavior of the estimated relative error is shown. Depending on the number of elements used for two load cases. Lines A_1-B_1 and A_2-B_2 represent the relationship between the errors and the number of elements for the optimal meshes of each load case. The slopes of these lines represent the maximum convergence speeds that can be achieved considering each case in isolation. Maximizing convergence speeds through the slope of a line in the context of mesh refinement offers several benefits, including efficient mesh refinement, improved solution accuracy, enhanced predictive capability, optimal resource utilization, and cost and time savings.

The optimal mesh, considering only load case 1, CC1, allows reducing the error for said case. Following line A1-B1, at point C1, there would be an error equal to the desired one with several elements equal to Nn (1). The mistake for CC2 has not been controlled and will correspond to a point D2 above the line A2-B2 of maximum convergence. Several factors can contribute to the instability of controlling CC2's error, including sensitivity to changes in the numerical solution, poor mesh quality, solution discontinuities, numerical instabilities, inaccurate modeling assumptions, and underestimated complexity. Addressing these issues requires refining the numerical solution method, improving mesh quality, reassessing modeling assumptions, and implementing more advanced error control strategies tailored to the specific challenges posed by CC2. As for the previous case, the result is analogous if it is refined considering CC2.

Figure 1. Maximum error convergence speed for various load cases

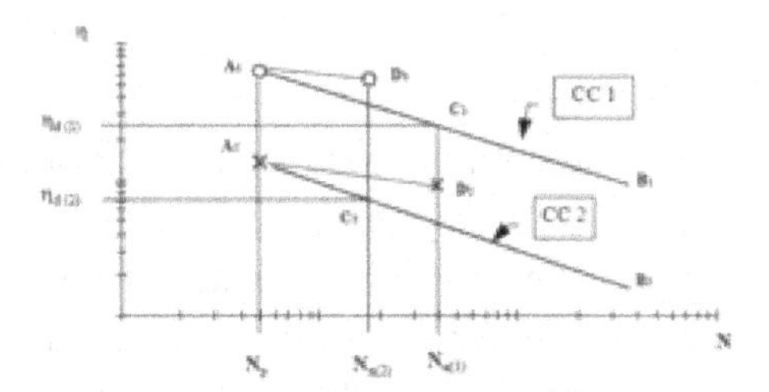

When more than two load cases are considered, it is possible to obtain the optimal mesh for the combination of all cases without considering some of them; that is, when refining the mesh considering some cases, the errors for the remaining cases are below the desired values. This can be better understood by analyzing the situation for two load cases.

When you want to determine the optimal mesh considering two load cases, two situations may arise depending on whether or not it coincides with the optimal mesh of one of the cases considered in isolation. Regular software updates and maintenance are essential for system security and stability. Updates include patches for known vulnerabilities, performance enhancements, bug fixes, and new features. Organizations can minimize security risks, prevent system failures, and maintain IT infrastructure health and reliability by staying up-to-date.

In the case of coincidence, it is possible that when refining considers only one load case, there will be an estimated error for the other case, less than or equal to the desired one. Figure 2 represents this situation. When refining with CC1, the optimal mesh is obtained at point C1, the number of elements under these conditions is point D2 represents Nn (1), and the estimated error for CC2, and although it has not been taken into account CC2 is counted for refinement, the condition that the error is less than or equal to the desired one is satisfied.

Note that if we try to achieve an estimated error in case two equal to $\eta_{d(2)}$ while maintaining the estimated error in case one is equal to $\eta_{d(2)}$, this would be achieved at points E_1 and E_2 by increasing the elements.

Figure 2. An optimal mesh for two cases that coincide with the optimal mesh of one of the cases is considered in isolation

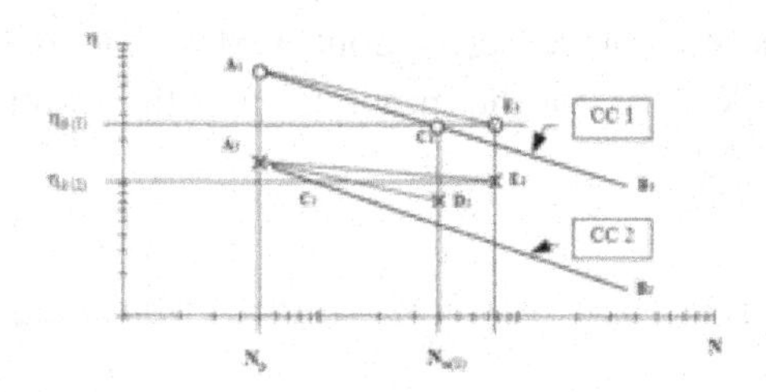

In the most general case, the optimal mesh for the combination does not coincide with the optimal meshes of any independently refined cases, figure 3. When combining multiple load cases, the optimal mesh may not coincide with the optimal meshes of any independently refined case due to different error characteristics, interactions between load cases, trade-offs and compromises, complexity of solution space, and system-level effects. This reflects the need for holistic approaches to mesh refinement that consider the combined impact of all load cases and strive for a balanced and optimized solution across the entire system. The mesh of Nn elements allows for estimating relative errors equal to those desired at points E_1 and E_2 for cases one and two, respectively. The slopes of lines A_1-E_1 and A_2-E_2 are less than the maximum of the errors in the cases considered in isolation, lines A_1-B_1 and A_2-B_2.

Figure 3. Optimal mesh for two load cases that does not coincide with the optimal meshes of the cases considered in isolation

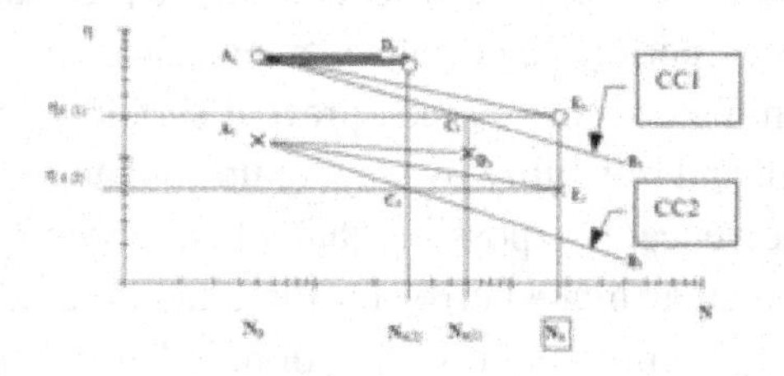

A criterion used in practice to define the refinement with several load cases is to calculate the local size ratio considering each case in isolation and then take the maximum value

$$r^{(e)} = max\left(r^{(e)}_{(i)}\right) \quad (26)$$

In this case, $r^{(e)}$ corresponds to the element size ratio for the new mesh. Even though this is a practical method that allows us to exceed the pre-established values, it has the disadvantage that these values go beyond the desired errors, unnecessarily increasing the computing time. This is mainly because a mesh with these characteristics would refine elements with size ratios larger than those necessary for the load case (Nochetto, R. H., et al., 2009).

A solution to this approach consists of limiting the size of the elements derived from equation (20) with a factor obtained from the desired values for the new mesh. Introducing this factor, equation (20) is expressed as

$$\|e_{es(i)}\|_n^2 = \sum_{e=1}^{N_p}\left(\|e_{es(i)}^{(e)}\|_p^2 \cdot \left(\rho_{(i)} r_{(i)}^{(e)}\right)^{-2c}\right) \tag{27}$$

where $\rho_{(i)}$ corresponds to each load case's size ratio correction factor. Solving for $\rho_{(i)}$, we have

$$\rho_{(i)} = \left[\frac{\sum_{e=1}^{N_p}\left(\|e_{es(i)}^{(e)}\|_p^2 \cdot \left(r_{(i)}^{(e)}\right)^{-2c}\right)}{\|e_{es(i)}\|_n^2}\right]^{-2c} \tag{28}$$

From the values obtained from equation (28), the maximum is taken, and the criterion to define the optimal mesh can be expressed as

$$r_{opt}^{(e)} = \rho_{max} \times r^{(e)} \tag{29}$$

Where $r_{opt}^{(e)}$ represents the optimal element size ratio, the new element sizes for the new mesh are obtained from

$$h_n^{(e)} = \frac{h_p^{(e)}}{r_{opt}^{(e)}} \tag{30}$$

Figure 4. Optimal mesh for the three load cases

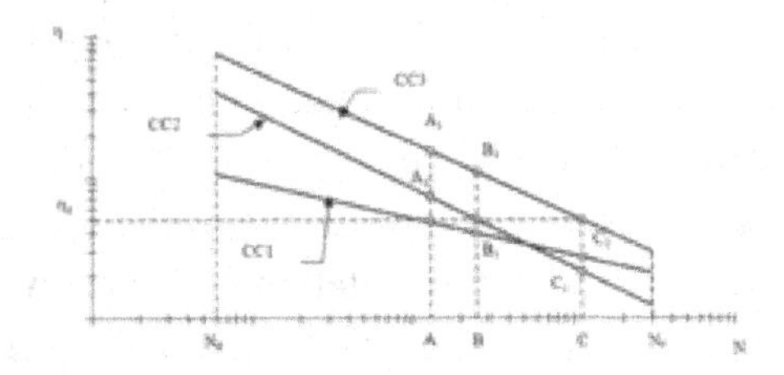

Figure (4) shows the relationship between the relative error and the number of elements in a refinement process for three load cases. The dashed lines N_p and N_r indicate the number of elements in the previous mesh and a mesh with an element size relationship obtained from equation (27). Lines A, B, and C are obtained by modifying the relationship of size with the correction factors $\rho_{(i)}$ corresponding to each load case. As can be seen, the values reached for several elements equal to Nr are below the desired value η_d; even when the desired value is reached, it is unnecessarily exceeded, increasing the computational time. Delays in project timelines, increased computational costs, limitations on simulations, and impacts on productivity and decision-making are all potential implications of increased computation time. In the case of the dashed line A, this is obtained by delimiting the size relationship with the correction factor $\rho_{(i)}$of load case 1, CC1; in this case, the relative errors of cases CC2 and CC3 point A2 and A3, respectively, do not reach the desired error. Dashed line A shows that refining the mesh to control errors specifically for load case 1 may neglect errors for other load cases. It prioritizes error control for load case 1 at the potential expense of error control for other load cases. For the dashed line B corresponding to the number of elements obtained with a size ratio limited by the correction factor of the load case CC2, even when the case CC1 manages to exceed the desired value at point B1, the same does not occur for the case CC3, point B3. In the dashed line C, obtained with the respective bounding factor, it is observed that case CC3 reaches the desired value and that cases CC1 and CC2 exceed the desired error with a more minor increase than that obtained with the number of elements N_r, points C1 and C2 respectively. Upgrading hardware components can boost processing speed and system performance. Key components to consider upgrading are CPU (faster CPU with more cores), RAM (more RAM), SSD (from HDD to SSD), GPU (dedicated GPU), Motherboard (newer model), Cooling System (better cooling), and PSU (higher wattage and more efficient). Upgrading these components can improve processing speed, multitasking, and system responsiveness.

(3) Development of the procedure

In the calculations preceding the refinement process, the errors in the previous mesh are estimated for each of the load cases (global and at the element level) of the solution of the FEM and the desired global errors for the new mesh are set. Several factors are used to determine the preferred global errors for a new mesh, including error tolerance levels, sensitivity analysis, mesh convergence studies, comparative analysis, error estimation techniques, and validation and verification. By analyzing these factors, researchers can establish the optimal error thresholds that balance solution accuracy with computational efficiency. The optimal mesh with the proposed formulation can be obtained through the following procedure.

i. Calculate the size of elements considering each load case.
ii. Determine the maximum size ratio for each element.
iii. Calculate the element size ratio correction factor for each load case.
iv. Determine the correction factor.
v. Calculate the optimal element size ratio.
vi. Determine the dimensions of the elements in the new mesh.

Machine learning algorithms can optimize system performance and efficiency in computational tasks by selecting suitable algorithms, predicting performance, optimizing resource allocation, detecting anomalies, and automating hyperparameter tuning. These techniques can help enhance computational efficiency and effectiveness.

NUMERICAL EXAMPLES

The proposed formulation was validated through several examples of satisfactory results. Numerical examples are crucial in validating and demonstrating the effectiveness of a proposed formulation. To ensure robustness and applicability, it is essential to consider utilizing diverse scenarios, benchmark problems, sensitivity analysis, convergence studies, complexity gradation, comparison with existing methods, and publication/documentation. By incorporating these considerations in additional numerical examples, researchers can effectively validate and emphasize the importance and relevance of the proposed formulation, bolstering its credibility and applicability in the field.

Figure 5 presents the geometry and data of one of these examples. It shows a structural component subjected to three load cases.

For the study, four h-adaptive processes have been defined. Three correspond to obtaining the desired error considering each load case in isolation, and the fourth uses the proposed refinement method. The estimated errors and effectiveness indices are previously calculated for all cases between successive meshes to obtain the mesh sequences.

In calculating the effectiveness index, the "exact" solution was estimated from an excellent mesh of triangular elements with an estimated error of less than 0.5%. The objective was set to achieve a 5% error for the three load cases, $\left(\eta_{d(1)} = \eta_{d(2)} = \eta_{d(3)}\right)$, in four stages. Researchers estimate the "exact" solution from a fine mesh of triangular elements using analytical solutions, high-resolution numerical solutions, mesh convergence studies, or interpolation from coarser mesh solutions. The estimated solution should always be validated against analytical solutions, benchmark problems, or experimental data to ensure its accuracy and reliability. The limitations and assumptions associated with each method should also be considered.

The analysis model uses linear triangular elements. It represents complex structures with a simple and efficient element type. They have geometric flexibility, are easy to mesh, are computationally efficient, exhibit good convergence properties, handle stress concentrations, and can be adapted for mesh refinement or increasing polynomial order. Analysis software packages widely support them and have been used in various engineering applications for accurate and efficient problem-solving. The initial mesh is illustrated in Figure 6.

As a result, Figures 7 to 10 present graphs of the relative estimated error and the effectiveness indices of the estimator as a function of the DOF. The thick line in the graphs (except Figure 10) represents the results of the load case controlling the refinement process, and the thin lines represent the results of the remaining cases. Fig. 11 illustrates, in descending order, the mesh sequences obtained for the three load cases and the proposed method. Each mesh represents a stage of the process.

Figure 5. Geometry of the example

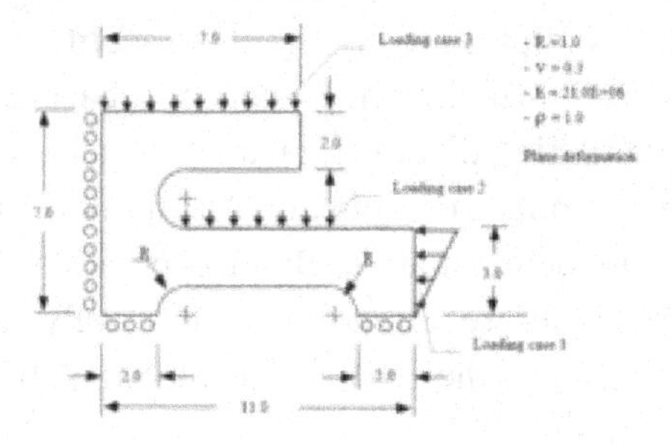

Figure 6. Initial mesh for linear triangular elements

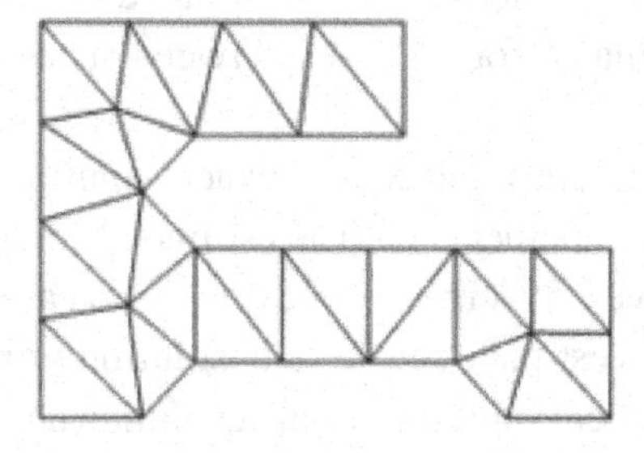

Figures 7, 8, and 9 show how the desired error is reached only by the load case considered for the refinement and not the remaining cases. The effectiveness index of the estimator tends to converge to unity for the case considered. A detail worth highlighting is obtained from Figure 7, where the refinement considering load case 1 almost manages to reach the objective for load case 2. A similar behavior, but on the contrary, is observed in Figure 8. This is explained by the fact that load cases 1 and 2 influence the same area of the structure (bottom). The mesh sequences for load cases 1 and 2 can result in varying stress distributions in the same structure area due to factors such as applied loads, boundary conditions, material properties, geometry, support conditions, loading sequences, and dynamic effects. Figure 11 illustrates how the refinement extends over this area in each stage, locating more elements towards the zone of influence of each load. The sequence of meshes for load case 3 clearly shows why the refinement, considering this load case, does not allow the objective to be reached for the remaining cases. Enhancing cyber security measures is essential to ensure computational systems' integrity, confidentiality, and availability. Proposed methods include intrusion detection systems, anomaly detection, threat intelligence, and vulnerability management. Existing methods include firewalls, encryption, access controls, authentication, incident response, and recovery. By combining these methods, organizations can enhance their cyber security posture and protect against cyber threats.

Figure 10 shows how the joint objective of all cases is achieved using the method proposed in this work. In the sequence of meshes corresponding to this procedure, it stands out how the meshes obtained in each stage are the product of the combination of all the areas affected by the loads, observing how the refinement simultaneously becomes denser in those areas where it is required. It's essential to consider some constraints, potential drawbacks, and future research possibilities to understand the proposed approach fully. The constraints include the need for significant computational resources, complex algorithms, and the accuracy and reliability of the results being influenced by underlying assumptions and simplifications. Potential drawbacks are overfitting and poor mesh quality due to excessive mesh refinement or error tolerance levels. Future research and development opportunities include developing more efficient and robust adaptive algorithms, exploring multiscale modeling techniques, and integrating uncertainty quantification techniques to assess the impact of input uncertainties, modeling assumptions, and parameter variations. Addressing these concerns can improve the proposed approach's performance and applicability in practical engineering applications.

Figure 7. Convergence and effectiveness index as a function of the DOF for load case 1

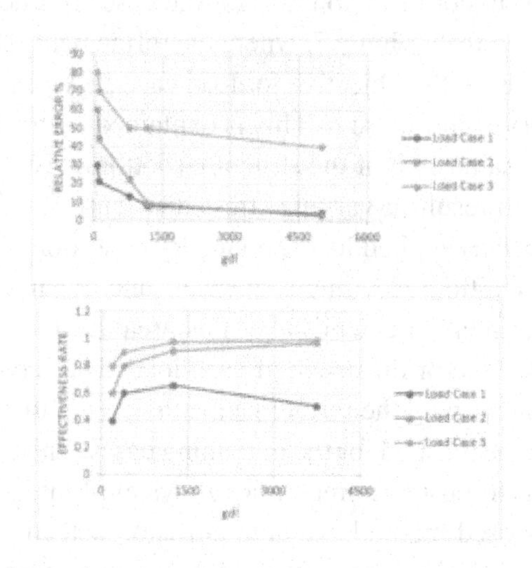

Figure 8. Convergence and effectiveness index as a function of the DOF for load case 2

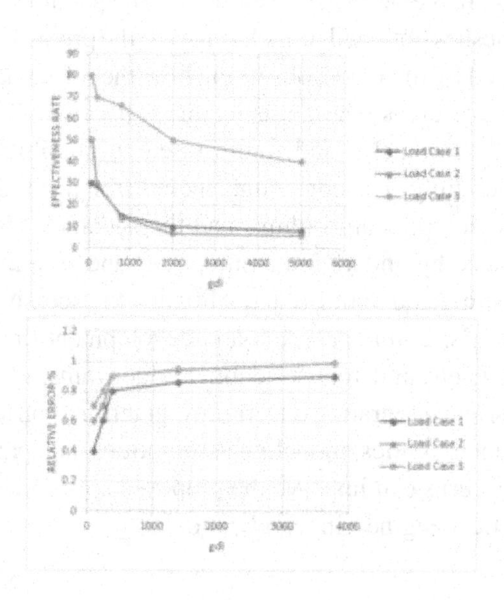

Figure 9. Convergence and effectiveness index as a function of the DOF for load case 3

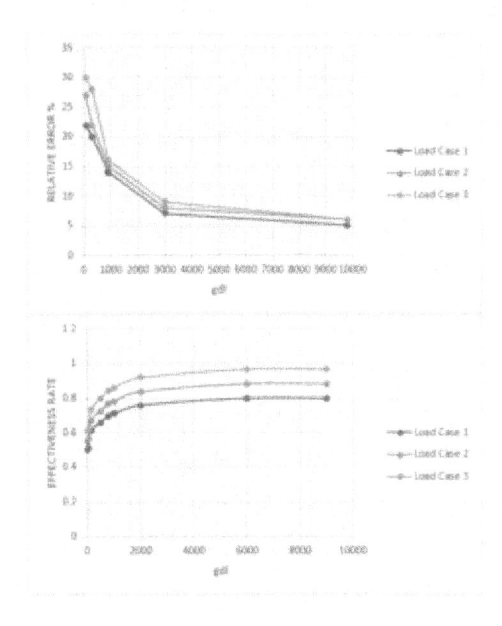

Figure 10. Convergence and effectiveness index based on the DOF generating optimal meshes

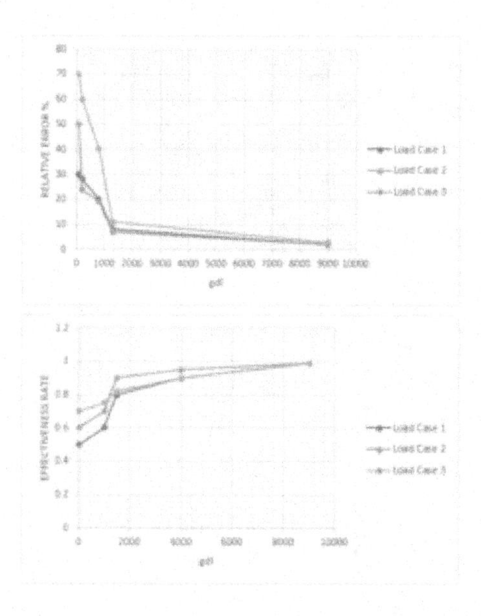

Figure 11. The sequence of H-Adaptive meshes for linear elements

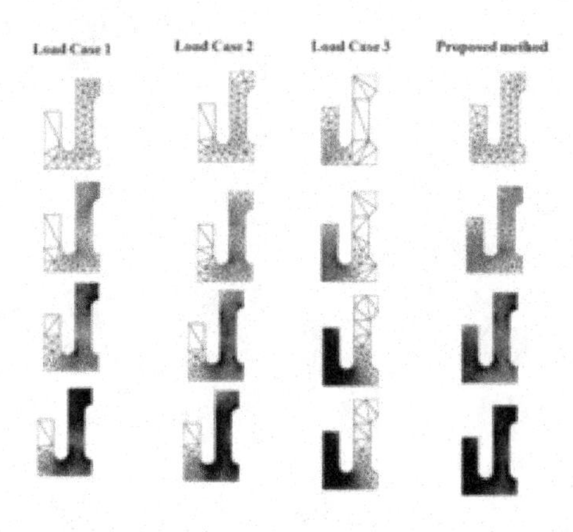

CONCLUSION

This work proposes a method for defining an h-adaptive procedure for solving two-dimensional static problems with multiple load cases. The proposed method was based on adapting the practical criterion of the minimum size of elements defined by the different load cases, subject to the concept of optimal mesh. This scheme allows for determining the optimal mesh in h-adaptive processes for multiple load problems with minimum elements, achieving an error less than or equal to a pre-established value for all cases. The procedure is computationally feasible and integrable within the h-adaptive process since it can be interleaved as a module between two sequential stages of the h-adaptive process. The proposed method has been numerically validated, one of which has been presented in the present work.

REFERENCES

Carson, H. A., Darmofal, D. L., Galbraith, M. C., & Allmaras, S. R. (2017). Analysis of output-based error estimation for finite element methods. *Applied Numerical Mathematics*, 118, 182–202. 10.1016/j.apnum.2017.03.004

Coorevits, P., & Bellenger, E. (2004). Alternative mesh optimality criteria for h-adaptive finite element method. *Finite Elements in Analysis and Design*, 40(9-10), 1195–1215. 10.1016/j.finel.2003.08.007

Erhunmwun, I. D., & Ikponmwosa, U. B. (2017). Review on finite element method. *Journal of Applied Science & Environmental Management*, 21(5), 999–1002. 10.4314/jasem.v21i5.30

Li, J., Guan, Y., Wang, G., Wang, G., Zhang, H., & Lin, J. (2020, June). A meshless method for topology optimization of structures under multiple load cases. In *Structures* (Vol. 25, pp. 173–179). Elsevier. 10.1016/j.istruc.2020.03.005

Mahomed, N., & Kekana, M. (1998). An error estimator for adaptive mesh refinement analysis based on strain energy equalisation. *Computational Mechanics*, 22(4), 355–366. 10.1007/s004660050367

Nochetto, R. H., Siebert, K. G., & Veeser, A. (2009). Theory of adaptive finite element methods: an introduction. In *Multiscale, Nonlinear and Adaptive Approximation: Dedicated to Wolfgang Dahmen on the Occasion of his 60th Birthday* (pp. 409-542). Springer Berlin Heidelberg. 10.1007/978-3-642-03413-8_12

Oñate, E., & Bugeda, G. (1993). A study of mesh optimality criteria in adaptive finite element analysis. *Engineering Computations, 10*(4), 307-321. .10.1108/eb023910

Pacheco, J. P. F. R., da Silva, J., & do Vale, J. L. (2020). *h-adaptative strategy proposal for Finite Element Method applied in structures imposed to multiple load cases*. Academic Press.

Chapter 5
Mathematical Foundations and Principles Behind These Methods

Noor Mohammed Kadhim
Wasit University, Iraq

Abdulsattar Abdullah Hamad
https://orcid.org/0000-0003-0497-5115
University of Samarra, Iraq

ABSTRACT

This work addresses the definition and historical introduction to numerical methods and the need for their application in engineering. Likewise, the basic concepts of accuracy, precision, convergence, and stability are established, as are the definitions of the different types of errors and how to quantify them. Finally, the Taylor polynomial is a resource for approximating mathematical functions. A subfield of mathematics known as "numerical analysis" uses iterative algorithms to produce numerical solutions to issues where analytical or symbolic mathematics is unable to solve or is inefficient. These algorithms are specifically referred to as numerical methods.

DOI: 10.4018/979-8-3693-3964-0.ch005

INTRODUCTION

A numerical method is an iterative mathematical process that aims to find the approximation to a specific solution with a certain previously determined error. Unlike the techniques of analytical mathematics, numerical methods require an approximation of the real solution to the problem, which is corrected through repetition. ´On a certain process, solutions must be yielded that are increasingly closer to the real value. Each correction of an initial value is known as an iteration.

There is no unanimity among experts on whether numerical analysis is a synonym for numerical methods. Some consider numerical methods to be processes with particular objectives that make up a more complex process, specifically the interpretation of the results, which they call numerical analysis.

It is difficult to take sides with either of the two previous positions, considering that the application of iterative processes is usually done to real problems with very specific design conditions, so it cannot be established a general rule for doing an analysis.

Historical Introduction of Numerical Methods

The history of numerical methods is the collection of mathematical events in which problems are solved without the use of analytical mathematics.

Some of the methods most used today were created long before the invention of the computer; Its application was exhausting and complicated because each iteration required a variety of arithmetic operations that were carried out by entire groups of calculators, obviously manually.

Everyone knowledgeable in the matter will agree that a computer performs a large number of operations in a very small interval; Super computers do it but in parallel. This capacity is what has given a sense of application to numerical methods.

Therefore, the history of numerical methods is parallel, at least since the middle of the 19th century, to the history of computing.

This is a list of events that have marked the history of numerical methods and the reader is recommended, according to their interest, to delve deeper into the topic that interests them; In particular, the work of Isaacson, (2014) offers a very broad overview of this matter.

NEED FOR THE APPLICATION OF NUMERICAL METHODS IN ENGINEERING

Numerical analysis is a branch of mathematics that, through iterative algorithms, obtains numerical solutions to problems in which symbolic (or analytical) mathematics is inefficient or cannot deliver a result. In particular, these algorithms are called numerical methods.

Numerical methods are generally composed of several finite steps that are executed logically, improving initial approximations to a certain quantity, such as the root of an equation, until a certain error level is met. This cyclic operation of value improvement is known as iteration.

Example, the latter existing in conjugate pairs. The commonly used solution method is a synthetic division (a numerical method). The student applies it as often as necessary to ensure that the remainder of the division is zero or close to zero.

However, this procedure could leave a diligent student dissatisfied because, even when there are mechanisms to choose an initial value of a root, a lot of time is spent improving this initial value. Additionally, it is complicated to obtain complex roots, which usually must be achieved through a change of variable and using the general formula for second-degree equations. Finally, this process only applies to polynomials; its application to transcendent equations is impossible.

Numerical analysis is a very efficient alternative for solving equations, both algebraic (polynomials) and transcendental, and it has a very important advantage over other types of methods: The repetition of instructions (iterations). This process allows for improving the values initially considered as a solution. Since it is always the same logical operation, it is pertinent to use computing resources to perform this task.

However, there must be clarity in that numerical analysis is not the panacea for solving mathematical problems; numerical methods provide approximations, which are subject to error. This error, although it can be as small as the calculation resources allow, is always present, and its management must be considered in developing the required solutions.

Various computer systems may be known to provide analytical solutions. These do not replace numerical methods; in fact, they complement the integral process of modelling physical systems, which is the fundamental element of engineering practice.

Engineering is the scientific discipline (Rionda, s.f.) that applies basic sciences to solve problems.

DEFINITION OF ERRORS

A frequent activity of the engineering professional consists of working with mathematical models' representative of a physical phenomenon. These models are mathematical abstractions that are far from accurately representing the phenomenon under study due mainly to the shortcomings and difficulties that humans still have in fully understanding nature.

As a consequence of this, there are differences between the results obtained experimentally and those emanating from the mathematical model.

The quantitative differences between the two models are called Errors.

Example: Let h be the height at which a body is located, g the acceleration constant of gravity and t the time the fall lasts, the mathematical model is defined as:

$$t = \sqrt{\frac{2h}{g}}$$

It is logical to think that when carrying out the calculations using the previous model, results will be obtained that will differ from the measurements that could be obtained during the development of the experiment.

Classification of Errors

The differences (errors) are multiple and of diverse nature, although they can be separated into two generic groups:

Errors that come from theoretical modelling (or mathematical abstraction) of the real phenomenon; These errors are called Model or Inherent Errors. Inherent errors are the product of factors intrinsic to nature, the environment and people themselves. Inherent errors are impossible to remedy although they can be minimized; Consequently, they cannot be quantified.

Two types of inherent errors are distinguished: Uncertainties refer to physical dimensions that can never be measured exactly due to the nature of the matter and the imperfections of the measuring instruments. The real mistakes are the situations that occur in the reading of measuring instruments or in the transfer of information and that are unnoticed by people; A clear example of these situations is the so-called workshop blindness.

The errors of the method are a product of the limitation in the representation and manipulation of numerical quantities used in the calculations necessary in the development of the mathematical model. It is noteworthy that calculating devices (such as calculators and computers) use and manipulate quantities imprecisely.

There are two main types of errors in the method: Truncation is caused by the impossibility of manipulating, by a computing instrument, an infinite number of terms or figures. Omitted terms or figures (which are infinite in number) introduce an error in the calculated results. Rounding occurs for the same reason as truncation but, unlike truncation, the omitted figures are considered in the resulting figure. This consideration is made by applying the following scheme to the least significant digit (*lsd*) of the figure to be rounded according to the following scheme:

1. If the *lsd* is greater than 5, the previous figure is increased by one unit.
2. If the *lsd* is less than 5, the previous figure is not modified.
3. If the *lsd* is equal to 5, the previous figure should be observed; If it is even, it is not modified, but on the contrary, if it is odd, it must be increased by one unit.

You may know a practical and popular version of symmetrical rounding in which consideration three is included in the first of this scheme. Finally, there are also schemes that allow the occurrence of these errors to be minimized. Likewise, it is important to highlight that the errors of the method can be quantified.

Error Quantification

Errors are quantified in two different ways:

1. Absolute Error. The absolute error is the absolute difference between a real value and an approximate one. It is given by the following formula:

$$E = \left| V_{Real} - V_{Aprox} \right|$$

The absolute error receives this name since it has the same dimensions as the variable under study.

The difference between the preference in the use of the two types of error consists precisely in the presence of the physical dimensions. Due to the measurement units used, the handling and perception of absolute error is often misleading or difficult to understand quickly. However, handling percentages (or relative values) is more natural and easier to understand. However, the use of these two types of errors is always subject to the objective of the activities carried out.

Considerations on Real Value (V_{Real}): The expressions that define absolute and relative errors require knowledge of the variable V_{Real}, which represents an ideal value that does not have any error. As might be expected, in reality it is impossible to determine this value.

A common practice in elementary error analyses is to consider as a real value the results obtained by the experimental measurement of the phenomena and the approximate values such as those provided by mathematical models (or *vice versa*). The reader has perceived that there is an error in both values, which is why neither of them can be considered a real value. In reality, both values are approximate values. For example, nominal values are considered real.

To achieve a consistent result, in practice the real value must be replaced by a value that is considered to have a smaller error. For example, in a measurement process, the nominal values cited in the specifications of the objects to be measured are usually used as the real value.

In the case of numerical analysis, given that the results are obtained from iterative processes that improve those initially obtained, it must be assumed that the last value obtained has a lower level of error than the previous value.

Given the above, the absolute and relative errors will be calculated as follows:

Absolut mistake:

$$E = | V_i - V_{i-1} |$$

Relative error: $e = | Vi - Vi - 1 |/Vi\, x100\ \%$

$$e = \frac{| V_i - V_{i-1} |}{V_i} \times 100\ \%$$

In both cases, V_i is the value of the last iteration i and V_{i-1} is the value of the previous iteration $i - 1$.

Magnitude of Errors Due to Truncation and Rounding

Unfortunately, specialized literature on error treatment is scarce, so it is nevertheless very important to know the magnitude of the errors that are committed, in this case, in the development of numerical methods. A study on errors widely spread among the community dedicated to the development of Numerical Analysis is the one developed by Daniel McCracken. The aforementioned study is focused on the management of numerical data on a computer and belongs to a historical moment in which computing resources were still very limited compared to those available in the beginning of the 21st century. In fact, McCracken's conclusions are still valid today.

An important contribution to the study of errors consists of the quantification of the magnitude of those that occur in the handling of data inherently in the use of floating-point arithmetic. Mc Craken concludes that the magnitudes of the errors committed by truncation are greater than those committed by the use of symmetrical rounding (McCracken and Dorn, 1984). Likewise, it is also concluded that the magnitude of the error due to symmetric rounding is independent of the quantity itself, being a product of the size of the mantissa used to make the calculations. The maximum absolute error due to symmetrical rounding is calculated through the expression:

$\frac{1}{2} \times 10^{-t+1}$ where t is the size of the mantissa

Example. Using a 3-digit mantissa, determine the maximum absolute error made in the following figures:

- 10.334
- 123293.967

It is observed that quantities 1 and 2 are very different in magnitude; However, the maximum absolute error present in each of them is the same.

It is important to establish that when performing calculations, it is not important to know the algebraic sign of the errors, the important thing is to know the difference between the work values, that is, their distance in absolute value. This must always be less than an allowable amount of error to consider the calculation valid. In engineering practice, this amount of allowable error is known as tolerance.

Tolerances are usually expressed in the form of percentages (relative errors) and are almost always focused on the number of significant figures that should be used in the approximation. It can be shown that if the following criterion is met, we can be sure that the result is correct to at least n significant figures:

$tol = (0.5 \times 10^{2-n})$ [%]

Example. Calculate the value of the function e^1 using the series:

$$e^x = \sum_{i=0}^{n} \frac{x^i}{i!} = 1 + x + \frac{x^2}{2!} + \frac{x^3}{3!} + \ldots$$

varying the number of terms of the series used and using five exact figures. For this example, the tolerance is $tol = 0.5 \cdot 10^{2-5} = 0.00050$. If the value obtained directly from a calculator is considered as the real value, the result is shown in the following table:

continued on following page

Table 1. Continued

Table 1. Errors in the calculation of infinite series

Term	Value	Error
1	1.01	1.735463
2	2.02	0.725463
3	2.525	0.220463
4	2.693337	0.052126
5	2.735413	0.01005
6	2.743837	0.001626
7	2.745241	0.000222

A second contribution of McCracken's study is the establishment of a process to measure the propagation of errors caused by the use of floating point arithmetic. Based on the establishment of the maximum absolute error committed and the arithmetic operation used, it is shown that in this type of process the order in which the operations are performed does modify the result.

Example. Add the following quantities, first in ascending order and then in descending order, considering a normalized four-digit mantissa as well as symmetrical rounding in each intermediate operation; On the other hand, perform the exact sum (with all possible digits on a calculator) and consider this value as exact. Calculate the relative error committed in each case.

0.2685×10^4

For the requested alternatives, the respective tables will show the normalized quantity as well as the subtotal, that is, the rounded sum in a normalized mantissa of size 4. The exact value, obtained through a calculator, is:

The procedure consists of normalizing the quantities (equalizing the exponent of the base ten in each quantity) and adding them in ascending or descending order, as the case may be; In the sum of each pair of quantities, the result is rounded, keeping the mantissa at the pre-established size.

Table 2. Descending sum

Quantity	Normalized Quantity	Subtotal

continued on following page

Table 2. Continued

Quantity	Normalized Quantity	Subtotal
		0.364210^4
		0.364310^4
		0.364410^4

Table 3. Ascending sum

Quantity	Normalized Quantity	Subtotal
0.1111	**0.1111**	
0.0053	0.0530	0.1614
0.9567	95.67	95.8341
0.2685	268.5	363.3341

Table 4. Comparison of results

exact value	3643.341		
Descending sum	0.3664	20.659	0.57%
Ascending sum	363.3341	10	0.27%

Table 2 shows the ascending sum and table three shows the descending sum. Finally, the results are included in table four.

Finally, this study yields three important conclusions that should be considered in the design of algorithms to execute numerical methods.

McCracken's conclusions are as follows:

1. When you are going to add and subtract numbers, you should always work with the smallest ones first.
2. If possible, avoid subtracting two approximately equal numbers. An expression containing such a subtraction can often be rewritten to avoid it.
3. An expression of the type can be rewritten in the form and as If there are approximately equal numbers inside the parentheses, perform the subtraction before the multiplication. This will avoid complicating the problem with additional rounding errors.
4. When none of the above rules apply, the number of arithmetic operations should be minimized. It remains a voluntary task to analyse these conclusions and verify the way in which they were obtained.

NUMERICAL APPROXIMATION AND ERRORS

An approximation is a value close to one considered real or true. This closeness, or difference, is known as an error.

Normally, the consideration of the validity of an approximation depends on the error level that the experimenter considers relevant depending on the context of the phenomenon under study. This implies that it must also be considered that magnitude should be a real value, which in the field of Engineering is rarely known, which forces the adoption of conventions.

Accuracy and precision In Engineering, accuracy is the ability of an instrument to measure a value close to the actual magnitude. Accuracy implies precision, but not the other way around. Accuracy and precision are not equivalent. Accuracy is the ability to get closer to the actual magnitude, and precision is the ability to generate similar results. Precision is achieved when an instrument repeats exact measurements when they are made consecutively. According to the definition of numerical approximation, accuracy is applied in numerical methods in terms of the ability of the method to generate a result very close to the real value; the closeness between accuracy and the concept of error is perceived. On the other hand, numerical methods through iterations generate increasingly accurate approximate values, that is, these iterations must be precise. Given the above, numerical methods should have accuracy and precision as qualities.

Errors in Numerical Calculations

By numerical method we mean a method for calculating the solution to a problem by performing only a finite sequence of arithmetic operations. Obtaining a numerical solution to a physical problem through the application of numerical methods does not always give us values as intended. The difference between the obtained (approximate) value and the exact value is called error. The aim is to give users of numerical methods an idea of the sources of errors, so that they can eliminate them, or at least control their value.

Let us then describe the process of determining the solution to a physical problem, using numerical methods.

- modelling: the mathematical model that describes the behavior of the physical problem is obtained;
- resolution: the numerical solution of the mathematical model is obtained through the application of numerical methods.

Source and Type of Errors

Solving a physical problem using a numerical method generally produces an approximate solution to the problem.

The introduction of errors in problem solving can be due to several factors.

Depending on their origin, we can consider the different types of errors:

initial errors of the problem (are external to the calculation process)

- -errors inherent to the mathematical model
- -errors inherent to the data errors associated with the use of numerical methods (occur in the calculation process)

Errors inherent in the model: A mathematical model rarely offers an accurate representation of real phenomena. In the vast majority of cases, they are just idealized models, since when studying natural phenomena, we are forced, as a general rule, to accept certain conditions that simplify the problem in order to make it tractable. The best models are those that include those characteristics of the real problem necessary to reduce errors at this stage to an acceptable level.

- **Errors inherent in data**: A mathematical model not only contains equations and relationships, it also contains data and parameters that are often measured experimentally, and therefore approximate. Approximations in the data can have great repercussions on the final result.
- **Rounding errors**: Whether calculations are carried out manually or obtained by computer or on a calculator, we are led to use finite precision arithmetic, that is, we can only take into account a finite number of digits. The error due to disregarding others and rounding the number is called rounding error.
- **Truncature errors**: Many equations have solutions that can only be constructed in the sense that an infinite process can be described as the limit of the solution in question. By definition, an infinite process cannot be completed, so it must be truncated after a certain finite number of operations. This replacement of an infinite process by a finite process results in a certain type of error called truncation error. In many cases, the truncation error is precisely the difference between the mathematical model and the numerical model.

There are 2 types of errors associated with the use of numerical methods to solve a problem on a computer or calculator: rounding errors and truncation errors. As a consequence of the occurrence of these errors, the numerical solutions obtained are, in general, approximate solutions.

Calculation of Values of Transcendent Functions

Rational functions (polynomials and quotients of polynomials) are the only ones whose values can be calculated using only a finite number of arithmetic operations. To numerically calculate values of a transcendent function, we can approximate it by a rational function.

Function Approximation

The approximation of functions is a central theme of numerical analysis. The reason for this is the occurrence of a large number of mathematical problems, involving functions, the solution of which is not possible (or very difficult) to determine by analytical methods. Examples of such problems are the calculation of the value of a definite integral when a primitive of the integral function is unknown, the determination of zeros of a function when there is no explicit formula to do so, the drawing of the graph of a function of which we know only some of its numerically or experimentally determined values. The strategy in developing numerical methods to solve these problems is based on replacing the given function with an approximating function, considered "simpler", whose behaviour is very similar to that of the given function.

Errors in Numerical Approximations

An issue inherent to a numerical approximation is the convergence of the chosen method.

Informal Definition 1. The convergence of an approximate method measures how close the numerical solution sequence approaches the exact solution, as the step size h decreases.

How small do the step sizes have to be to ensure a certain level of accuracy? In some cases, if the step size is too small, the method may even lose accuracy.

The following content can be found in (Boyce and DiPrima, 1970). There are two fundamental sources of error:

Truncation error: Assuming we are working with absolute precision (infinite number of decimal places).

The global truncation error is:

where) is the exact solution and is the approximate solution. Causes:

- At each step, the approximate formula is used to find
- The input data at each step is not correct, that is,

If we assume that then the error made at each step is due to the use of an approximate formula. And the local truncation error is defined as:

Rounding error: When performing calculations with a finite number of digits, a rounding error is generated.

where is the value actually computed by the numerical method (e.g. on a computer).

Absolute error value: The absolute value of the total error in calculating is given by:

Using the triangular inequality:

The total error is limited by the sum of the absolute values of the truncation and rounding errors.

CONVERGENCE AND STABILITY OF A NUMERICAL METHOD

Mathematically, convergence is the property of some sequences and series of progressively tending to a limit, in such a way, if this limit exists, it is said that the sequence or series converges. In an analogous way, if a numerical method in its iterative operation provides us with increasingly closer approximations to the desired value, the method is said to converge. Convergence is measured through errors; If the error between two successive approximations is reduced, the method converges; It must be fulfilled that:

That is to say, the difference must be less than the () the difference .

A system (or a process) is said to be stable if small variations in the input or excitation correspond to small variations in the output or response. The stability of a numerical method has to do with the way in which numerical errors propagate throughout the algorithm. When a method converges, the most desirable thing is that in the results obtained, the error levels are reduced as quickly as possible. However, it happens that during the operation of the algorithm, either due to the handling of the numerical data or due to the nature of the mathematical model with which one is working, the errors between approximations do not decrease. progressively, but even increase at some stage of the process and then reduce, showing random behavior.

The robustness of a numerical method lies in its convergence and stability. Methods can be used whose convergence test indicates the relevance of their use, but during their application unstable results are obtained that affect the number of iterations and, consequently, the time invested in the solution. The ideal is methods that, while being convergent, are stable.

APPROXIMATION OF FUNCTIONS BY MEANS OF POLYNOMIALS

Particularly in the handling of transcendent functions, analytical problem solving can be difficult and complicated; even this situation could occur in the scope of the numerical solution. When this occurs, a possible solution is to use an approximate representation of the function through simpler functions. Some of these approaches are:

Periodic functions (sines and cosines) through Fourier series Segment the function through a sequence of straight lines

The Taylor Series

The expansion in Taylor series seeks to obtain an approximation to through a polynomial of the form: (1)

in the neighbourhood of the point x = x0 for its first n derivatives. Due to the above, it is required that f (x) has n – 1 derivatives in the interval a ≤ x ≤ b, that is, that:

(2)

It is necessary to determine the coefficients of the polynomial (1) and the derivatives valued at =0:

(3)

For

Substituting (2) and (3) into (1):

(4)

Expression (6) represents the McLaurin Series.

In a particular case, it is likely that the polynomial P(X) is required be equal to the function f(x) at a point X different from zero, that is, X = a 0, we proceed in the same way

(5)

This consideration generates a growth of the abscissa, so the general expression is:

(6)

Equation (5) is known as a Taylor polynomial of degree n for the function at the point .

Example. Compute the Taylor polynomials of degrees 1 and 3 for /2.

Let:

(7)

and

(8)

The derivatives valued at π:

(9)

Substituting (9) in (7) and (8):

Figure (1) graphically shows each of the approximations to

Taylor Polynomial Remainder

We must not lose sight of the fact that the Taylor polynomial is an approximation to the function entails an error that is not usually considered but that depending on its order could become significant. In this way, to the expression:

(10)

It is known as Taylor's formula with remainder. To calculate equation (6.1) is evaluated for various orders:

First order:

Figure 1. Approximations to the function

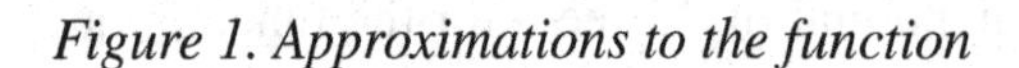

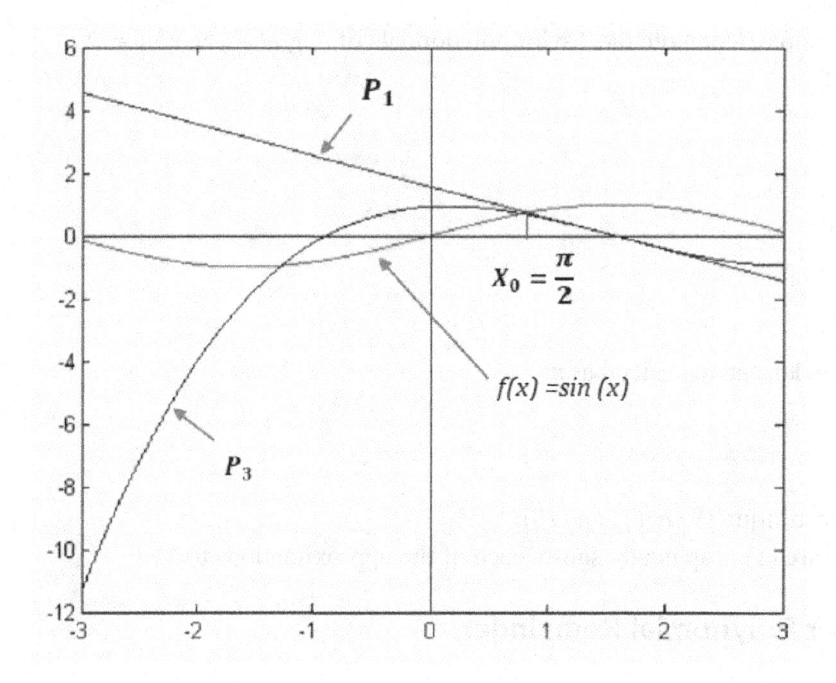

Solving for

(11)

Equation (11) can be expressed in integral form:

(12)

Integrating by parts:

(13)

For a second order the result is:

(14)

And for the nth order:

(15)

Equation (15) is the error made when approximating the function with a Taylor polynomial of degree

Estimation of the Error of the Taylor Approximation

Since the Taylor approximation represents a series with an infinite number of terms, it is not possible to find an exact value for so it is necessary to make some considerations: suppose that m and M are the minimum and maximum values, respectively, that the function acquires in the interval Substituting these assumptions into equation (15):

Both expressions are the error limits, that is:

Setting the values of and is a complicated problem. In real applications, a practical criterion is taken that consists of evaluating the term of the series at some point of interest that is in the neighborhood of = a. For example, we wish to estimate the error made when approximating , for = 0, through a sixth-order polynomial. The Taylor polynomial of is widely known:

The next term of the series to estimate the error is: $E_6 \leq$. It is defined as an inequality because the real value of the error will be closer to the pivot point, in this case = 0,

so, the error will be smaller. This is verified with the following values of :

As the value of approaches the point at which the polynomial was defined (in this case = 0) the error decreases.

CONCLUSION

The basic concepts of accuracy, precision, convergence and stability are established, as are the definitions of the different types of errors and how to quantify them. Finally, the Taylor polynomial is a resource for approximating mathematical functions.

REFERENCES

Arnold, M., Burgermeister, B., Führer, C., Hippmann, G., & Rill, G. (2011). Numerical methods in vehicle system dynamics: State of the art and current developments. *Vehicle System Dynamics*, 49(7), 1159–1207. 10.1080/00423114.2011.582953

Borras, H., Duran, R., & Iriarte, R. (1984). *Notes on numerical methods*. UNAM Engineering Faculty.

Boutayeb, A., & Abdelaziz, C. (2007). A mini-review of numerical methods for high-order problems. *International Journal of Computer Mathematics*, 84(4), 563–579. 10.1080/00207160701242250

Boyce, W. E. & DiPrima, R. C. (1970). *Elemental Differential Equations and Boundary Value Problems*. J. Wiley.

Finkelshtein, A. M. (n.d.). The numerical analysis in the last 24 years. *Magazine of the Spanish Society of History of Sciences and Techniques*, 26, 919–928.

Isaacson, W. (2014). The innovators: How a group of hackers, geniuses, and geeks created the digital revolution. *Journal of Multidisciplinary Research*, 7(1), 111.

McCracken, D., & Dorn, W. (1984). *Numerical methods and Fortran programming* (Limusa, Ed.). Academic Press.

Milici, C., Draganescu, G., & Machado, T. (2019). Numerical Methods. 10.1007/978-3-030-00895-6_

Rionda, S. B. (n.d.). *Examples of application of numerical methods to engineering problems*. A.C. Mathematics Research Center.

Russell, A. (2016, January-March). The innovators: How a group of hackers, geniuses, and geeks created the digital revolution (Isaacson, W.; 2014) [book review]. *IEEE Annals of the History of Computing*, 38(1), 94–c3. 10.1109/MAHC.2016.8

Sandoval-Hernandez, M. A., Vazquez-Leal, H., Filobello-Nino, U., & Hernandez-Martinez, L. (2019). New handy and accurate approximation for the Gaussian integrals with applications to science and engineering. *Open Mathematics*, 17(1), 1774–1793. 10.1515/math-2019-0131

Chapter 6
Parallel Computing Techniques

Alnoman Mundher Tayyeh
Mangalore University, Iraq

Akram H. Shather
Sulaimani Polytechnic University, Iraq

Saja Sumiea Anaz
Mangalore University, Iraq

Firas T. Jasim
Al-Dour Technical Institute, Northern Technical University, Iraq

ABSTRACT

Meshless methods are numerical methods for solving partial differential equations. The connectivity relationships between the discretized nodes do not need to be explicitly established. Instead, a collection of nodes is distributed over the problem domain. This represents an advantage when it is necessary to generate a new mesh several times, as in the case of moving structures. However, this advantage has its price, as this class of methods consumes more computing time than methods such as finite element methods. Thus, for more complex problems that require "remesh" such a moving electrical machine models, this work aims to apply parallel programming techniques to accelerate the execution of computational codes that solve mathematical models of electromagnetism based on meshless methods. Programming with threads, in this case open multi-processing, was used to parallelize the most time-consuming parts of the code of an electromagnetic model of an induction machine based on the Element-Free Galerkin meshless method.

DOI: 10.4018/979-8-3693-3964-0.ch006

INTRODUCTION

Parallel programming has been applied in several areas of knowledge, as it contributes by reducing the time spent processing certain algorithms and the amount spent on acquiring high-performance computing equipment (hardware) (Navarro et al., 2013).

The advancement of development practices using parallel programming was favored by the constant need to improve computational performance; in which the implementation of numerical simulation scenarios was necessary, but it was essential that this occurred within a more precise simulation scope. Currently, multi-core processors are becoming increasingly common in computers, cell phones and other electronic devices. Consequently, increasing the number of cores increases the number of threads (processing flows) available to work on, which results in a greater range of processing on the same equipment. Parallel computing has helped in solving several problems in diverse areas such as: engineering with general problems of numerical integration, calculation of sparse matrices, decomposition into singular values, even in biomedicine, as described by Drmač (2020), which uses this process to determine the three-dimensional structure of viruses and calculate the solvent-accessible surface area of proteins. Another scenario that applies to parallel computing is that of experimental models, such that it is possible to work with models on a reduced scale, work with non-destructive techniques, directly reduce the cost of simulations in laboratories, in addition to also reducing in specific cases biological and/or chemical risks.

Currently, parallel computing is applied in different segments, as shown in Figure 1. The next segment is research, responsible for 20.6% and subsequently, academic subjects, 16.6%. The remaining 6.4% is distributed between Governments, commerce and other unclassified segments (Figure 1).

Figure 1. Distribution by processing segment

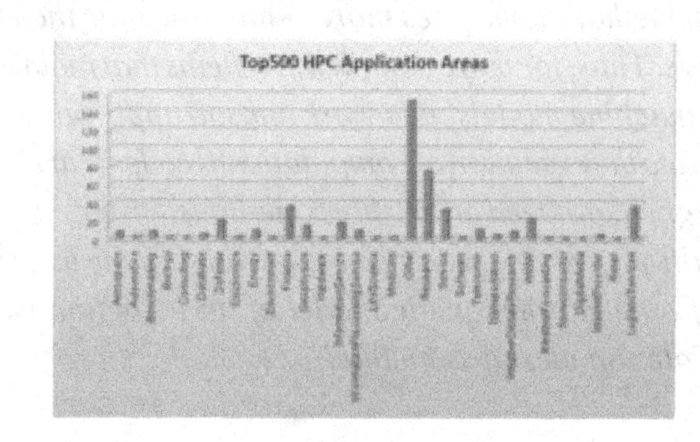

As a way of executing the growing processing demand, some architectures and technologies were created with the aim of enabling the concept, making the use of parallel programming viable. Among these architectures, Open Multi Processing (OpenMP) emerged, created in 1997; considered a thread-based library, which was built with the aim of optimizing the processing of algorithms on machines that have multiple cores; This process takes place via shared memory or multiprocessors.

Meshless are part of a group of numerical methods commonly applied to solving partial differential equations (PDE) (Li, 2005). These methods have as their main characteristic the fact that it becomes unnecessary to use meshes such as those used in finite element methods (FEM) (Idelsohn et al., 2003), for example. This characteristic is very important when it comes to modeling mobile structures in which its variation consists of reconstituting a new mesh for the integration process with each new formalization of the geometric model. Regardless of the numerical method, the integration procedure is computationally expensive, making it interesting to use parallel programming with the aim of reducing integration time.

This work will also present the use of parallel programming applied directly to the solvers through the PARDISO library (Schenk & Gärtner, 2002). This is a thread-safe library (manipulates shared data structure ensuring safe execution across multiple threads at the same time) with high performance, robust and highly efficient memory usage; mainly used for solving symmetric and non-symmetric sparse linear systems through the process of shared memory and distributed among other processors.

MESHLESS METHODS

The computational modeling of a physical phenomenon usually involves surveying the equations that describe its behavior, with the majority of these models being based on partial differential equations or PDEs.

Having collected the equations, a natural step in this process is the solution of these EDPs through a numerical method. The correct choice of method becomes a decisive factor in issues related to precision, stability and the computational cost involved.

The Finite Element Method (FEM), along with the Finite Difference Method (FDM) and the Finite Volume Method (FVM), are notable mesh-based methods for solving mathematical models. However, some characteristics observed in certain types of problems motivated the search for new numerical techniques in an attempt to resolve such issues. Consider, for example, problems where boundaries change over time, such as a moving electrical machine, or when cracks in a surface continually increase. Another example would be a structure that deforms as time evolves. In these cases, when a method such as finite elements is applied, what normally happens is a marked deformation in the elements, generating a loss of precision. To overcome this type of situation, a new mesh is built as these borders move. This process, in addition to being non-trivial, can be computationally expensive, especially when dealing with three-dimensional geometries. For problems of this nature, a new class of method has been developed, the so-called Meshless Methods.

This class of methods originated with the Smoothed Particle Hydrodynamics (SPH) or Smoothed Particle Hydrodynamics Method, which dates back to 1977. However, only since 1990 have such methods experienced strong development, mainly driven by the evolution of digital computer technology. Since then, several meshless methods (Nguyen et al., 2008) have emerged.

Initially, meshless methods had as their main applications problems linked to computational mechanics, an area that is still intensively explored. As for its application in electromagnetism, although there are records of work in 1992, with Marèchal as a precursor, its effective use is much more recent, dating back to the mid-1990s (Guimarães et al., 2007). Currently, these applications have reached significant numbers with publications in various journals and annals of specialized conferences (Guimarães et al., 2007; Nguyen et al., 2008).

The main characteristic that differentiates these methods from others, such as FEM, is basically that in meshless methods, a number of nodes are distributed in the domain under study, with no connection or relationship being pre-established between the nodes. same. These nodes will constitute the place where the unknowns must be determined. As will be shown in this work, this feature will facilitate the modelling of mobile structures.

ELEMENT-FREE GALERKIN METHOD (EFGM)

EFGM is a meshless method whose main characteristics are: 1) Moving least squares (MLS) are generally used to construct the shape function; 2) The Galerkin Method is used to arrive at the final system of equations; 3) A grid of integration cells is placed throughout the domain to perform the integrals present in the formulation.

MLS uses a local approximation function given by:

$$u^h(x) = \sum_{j=1}^{m} p_j(x) a_j(x) = \left\{ \underbrace{1 x y \cdots p_m(x)}_{p^T(x)} \right\} \left\{ \underbrace{\begin{matrix} a^{1(x)} \\ \vdots \\ a_{m(x)} \end{matrix}}_{a} \right\} \quad (1)$$

such that$X^T = [x, y]$*and* $p(x)$is a vector formed by monomials. The unknown parameters $a(x)$ are determined by minimizing the discrete norm in L_2 given by:

$$J * \sum_{I=1}^{n} w(x - x_I)\left[P^T(x_I)a\left(x\right) U_I\right]^2 \qquad (2)$$

such that w is the weight function. Minimizing equation (2) leads to:

$$a(x) = A^{-1}(x)B(x)u \qquad (3)$$

such that

$$a(x) = \sum_{I=1}^{n} w(x - x_I)\left[p(x_I) P^T(x_I)\right. \qquad (4)$$

and

$$B(x) = \left[w(x - x_I)\left[p(x_I)\ldots w\left(x - x_n\right)p(x_n)\right.\right. \qquad (5)$$

replacing (3) into (1)

$$\mathrm{u}^{\mathrm{h}}(\mathrm{x}) = \mathrm{p}^{\mathrm{T}}(\mathrm{x})\,\mathrm{A}^{-1}(\mathrm{x})\mathrm{B}(\mathrm{x})\mathrm{u} \qquad (6)$$

or

$$\mathrm{u}^{\mathrm{h}}(\mathrm{x}) = \sum_{I=1}^{n} \varphi_I(x)\, u_I \qquad (7)$$

where,

$$\varphi_I(x) = \mathrm{p}^{\mathrm{T}}\,\mathrm{A}^{-1}B_I \qquad (8)$$

is the shape function using MLS.

This shape function is used to approximate the unknown of the weak form of the problem in a similar way to the FEM. Consider the following two-dimensional problem in the domain Ω delimited by $\Gamma = \Gamma_i\, U\, \Gamma_u$

$$-\nabla.\,(\mathrm{k}\ \ \nabla\ \ \mathrm{u}) = \mathrm{b}\ \mathrm{em}\ \Omega \qquad (9)$$

$$-k\frac{du}{dn} = \bar{t}\ em\ \Gamma_t \qquad (10)$$

$$u = \bar{u} \ em \ \Gamma_u \quad (11)$$

Due to the fact that the function developed through MLS does not meet the Kronecker delta, it is necessary to impose essential boundary conditions through Lagrange multipliers.

Note that H^1 and H^0 denote Hilbert space of degree one and zero, respectively. In order to impose the Dirichlet boundary conditions, it is necessary to replace the MLS with the Interpolating Moving Least Squares (IMLS) in the terms in (12), thus discarding the Lagrange multipliers, since the shape functions of the IMLS have properties of the Kronecker delta function. Model similar to that adopted in FEM.

The final system of equations, after discarding the terms related to Lagrange multipliers, making use of IMLS elements, can be obtained by substituting the discrete form of the test function in (5):

$$KU = F \, (13) \quad (13)$$

$$K_u = \int_{\Omega} T_I^T \, k \, T_J d\Omega \quad (14)$$

$$T_I = \begin{bmatrix} \phi_{I,x} \\ \phi_{I,y} \end{bmatrix} (15) \quad (15)$$

$$F_1 = \int_{\Omega} \phi_I \mathrm{b} d\Omega - \int_{\Omega} \phi_I \bar{t} d\Gamma \quad (16)$$

MODEL TO BE PARALLELIZED

The model adopted for study in this work consists of a 2HP, four-pole, three-phase induction machine, with current supply and squirrel cage rotor and the iron in the machine was considered a material with linear response. Figure 2 represents the configuration used in two dimensions with the domain reduced to a quarter of the machine.

Figure 2. Geometry of the Induction Machine

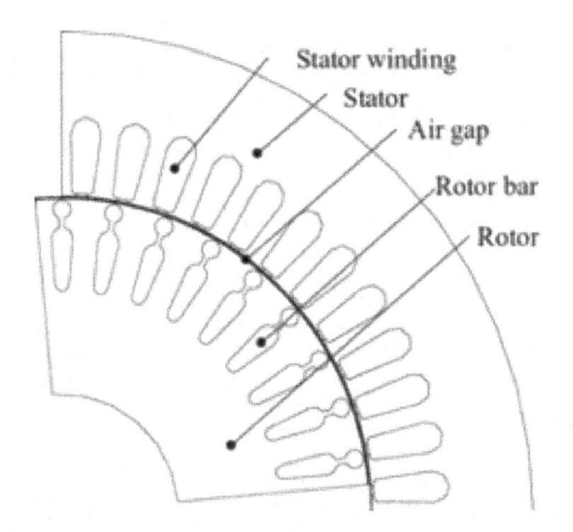

The following equations represent the electrical circuit field coupling model of the machine:

$$\nabla . v \ \nabla \ A - J_s = 0 \tag{17}$$

on the stator:

$$\nabla . v \ \nabla \ A - \sigma \frac{\partial A}{\partial t} + \sigma \frac{U_t}{l} \ = \ 1 \tag{18}$$

on the rotor, with:

$$U_t = R_t I_t + R_t \int_{S_t} \sigma \frac{\partial A}{\partial t} \, ds \tag{19}$$

where ***A*** is the magnetic vector potential, υ is the magnetic reluctance, σ is the electrical conductivity and J_s is the current density imposed on the stator. U_t is the voltage at the terminals of the rotor bars and I_t is the current passing through it. S_t is the sectional area of each rotor conductor, 1 its length and $R = (\sigma S_t)^{-1}$ refers to the d.c resistance. of the rotor.

The rotor circuit equations can be expressed by (20), in which Kirchhoff's Law was used.

$$C_1^T C_1 U_t + C_2 I_t \ = \ 0 \tag{20}$$

on what

$$I = \left[I_1, I_2, I_3 \ldots I_n\right]^T \quad (21)$$

$$C_1 = \quad (22)$$

$$C_2 = \quad (23)$$

$$U_t = \left[U_{t1} U_{t2} U_{t3} \ldots U_{tm}\right]^T \quad (24)$$

considering that the relationship $\boldsymbol{I}$ and $\boldsymbol{I}_t$ can be expressed by

$$I_t = C_1^T I \quad (25)$$

according to Kirchhoff's law. The EFGM is then applied to equations (17) to (19), resulting in:

$$KA = N\frac{d}{dt}A - PU_t - J = 0 \quad (26)$$

on what:

$$K(k,j) = \int_\Omega \nabla \phi_k^T v \ \nabla \ \phi_j d\Omega \quad (28)$$

$$N(k,j) = \int_\Omega \sigma \phi_k^T \phi_j d\Omega \quad (29)$$

$$P(k,j) = \int_\Omega \frac{\sigma_j}{l} \phi_k d\Omega \quad (30)$$

$$Q(k,j) = v \quad (31)$$

The elements of vector $\boldsymbol{J}$ are expressed by:

$$J(k) = \int_\Omega J_s(t) \phi_k d\Omega \quad (32)$$

After applying Euler's finite difference scheme in (26) and (27), using (30) and making use of the auxiliary matrices $\boldsymbol{C}_1$ (22) $\boldsymbol{C}_2$ (23) necessary in the circuit equations, and $\boldsymbol{C}_3$ given by (34),

$$C_3 = \tag{33}$$

the final system of equations is obtained:

$$\begin{bmatrix} K + \frac{N}{\Delta t} & -P & 0 \\ \frac{Q}{\Delta t} & C_3 & R \\ 0 & C_1^T C_1 C_2 & \end{bmatrix} \begin{bmatrix} A(t + \Delta t + T \\ U_t \\ I_t \end{bmatrix} = \begin{bmatrix} \frac{NA(t)}{\Delta t} + J\left(t + \Delta t + T\right. \\ \frac{QA(t)}{\Delta t} \\ 0 \end{bmatrix} \tag{34}$$

PARALLELIZATION STRATEGY

As described in the previous sections, the proposed problem is complex and works with the manipulation of mass data, in addition to the data being composed of values with decimal precision, which means that parallelism decisions for the problem must be taken carefully so that it does not occur rounding error.

It is notable that the repetition loop referring to the integration cells presents the greatest consumption of the algorithm's time.

Table 1. Percentage of time consumption of the main modules of the problem considering time = 1

Region	Percentage of Time Consumption
Initial Settings and Data Entry	3.17%
Repeat Loop referring to integration cells	55.65%
Other operations such as: 1) determination of boundary conditions; 2) Determination of the support domain; 3) Calculation of the shape function; etc.	23.04%
Solver	18.40%

Based on the cost of the integration loop presented in Table 1, a sub-flow was modelled, which can be seen in Figure 3.

Figure 3. Subflow referring to one of the candidate sections for parallelization

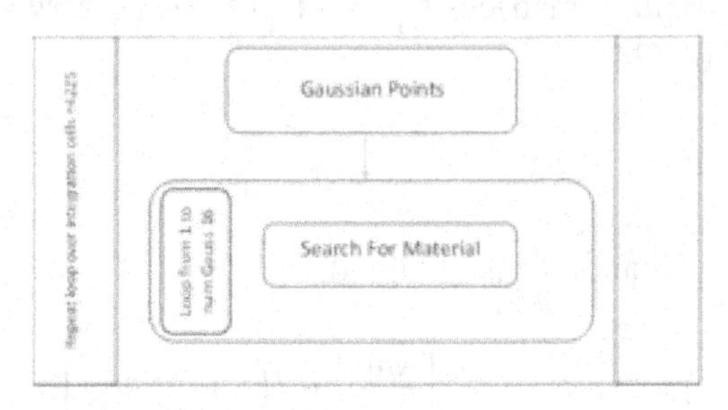

This section of code aims to perform a repetition loop over the integration cells where the Gauss points are located. These cells cover the entire domain and allow the integration operations found in the formulation to be carried out. It is worth highlighting that the subroutine search Material is present in this subflow.

serach Material is a subroutine developed in Fortran F90 with the aim of locating the type of material referring to the current integration point. As data input, this subroutine needs to receive the Gaussian points and returns the type of material. Internally, the process to locate the material is done by first identifying whether the point is within a micro region with defined material. If found, the search method is triggered to identify the type of material through a kd-tree type data structure used to store information during program execution.

OpenMP

OpenMP is an API (Application Programming Interface) for parallel programming used through the use of shared memory. The suffix "MP" in OpenMP describes the idea of multi-processing, which refers to parallel computing in shared memory (Ge et al., 2009).

Through OpenMP, the parallelization of code blocks can be carried out in a very simple and direct way, in addition to being very precise, focusing directly on the specific point to be parallelized.

OpenMP was developed by a group of programmers and computer scientists who believed that it was too complex to work with large-scale parallelization through the model known to date, Pthreads. To identify the use of OpenMP directives, the section to be parallelized always begins with the character # for the C/C++ languages and !$ for the Fortran language. Right after these initial characters, OpenMP implementations are encoded. It is very important to note that before using the OpenMP API, you must import the library into the code using the use omp_lib command.

It is worth noting that the variation in the thread number may vary due to its availability for execution at the time the code is started.

Table 2. Definition of OpenMP commands for the Hello.f90 Program

Code	Line in code	Function
useomp_lib	3	Import of the library and its dependencies required for use within the code block
	9	Definition of variables that now have their private context and are no longer shared with other threads at their runtime.
	19	Defining the End of a Parallelized Code Snippet

Solver PARDISO

PARDISO is a set of thread-safe, high-performance, robust, memory-efficient libraries focused on solving both symmetric and non-symmetric sparse linear systems. The resolution of these systems occurs via shared memory and distributed memory between multiprocessors.

As described by the website that maintains the project, the solver has been used in several international universities and scientific laboratories since its origins in 2004. PARDISO works with a wide range of combination variations during its solution process, such as: non-symmetrical, symmetrical matrices, with complex or real precisions, positive definite or indefinite (Schenk & Gärtner, 2002). The combination of left and right-looging via BLAS level 3 makes the solver's parallel augmentation and sequencing have a high performance.

Figure 4 shows the possible types of matrices that can be solved with the PARDISO solver.

Figure 4. Sparse matrix models that can be solved with the PARDISO solver

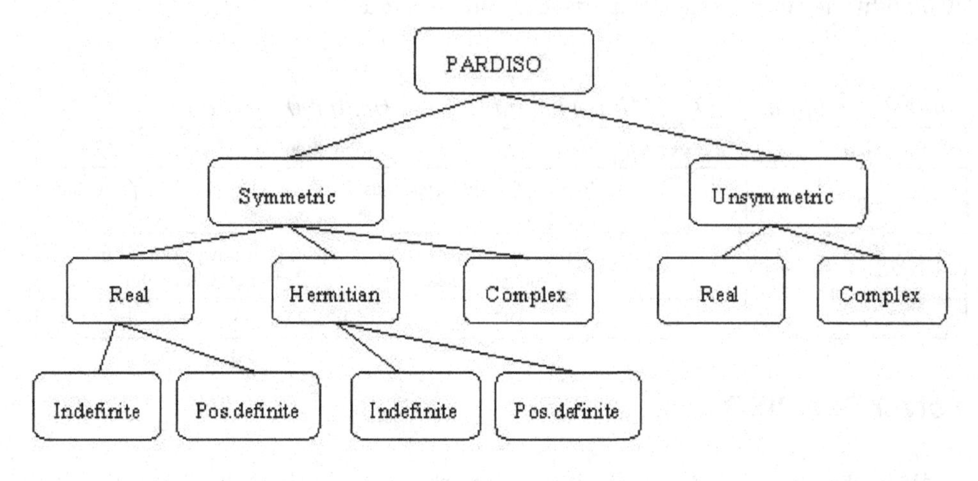

RESULTS

The presentation of results of this work is subdivided into 2 categories:

- Repeat loop over integration cells
- PADISO solver

For the first category, we work with the implementation of OpenMP in the most computationally expensive repetition loop for problems in electromagnetism related to the induction machine problem presented. Table 3 presents the relative consumption percentage of each step called most important in the serial algorithm.

Table 3. Percentage of time consumption representation in the serial algorithm

Region	Time Consumption Percentage
Initial Settings and Data Entry	3.17%
Loop for integration cells	55.65%
Other operations such as: 1) determination of border conditions; 2) Determination of the support domain; 3) Calculation of the form function; and so on.	23.04%
Solver	18.14%

With the strategy of parallelizing the repetition loop over the integration cells, the consumption percentage went from 55.65% to 4.37%, increasing the time for the Solver as shown in Table 3.

Table 4. Percentage of time consumption representation in the parallel algorithm only in the repetition loop over the integration cells

Region	Time Consumption Percentage
Initial Settings and Data Entry	2.14%
Loop for integration cells	4.37%
Other operations such as: 1) determination of border conditions; 2) Determination of the support domain; 3) Calculation of the form function; and so on.	16.48%
Solver	77.01%

With the aim of equalizing representation between the execution stages, the third stage was taken as an objective. This step consisted of parallelizing Solver through PARDISO.

Figure 5. PARDISO solver parallelization process ranging from one to six threads

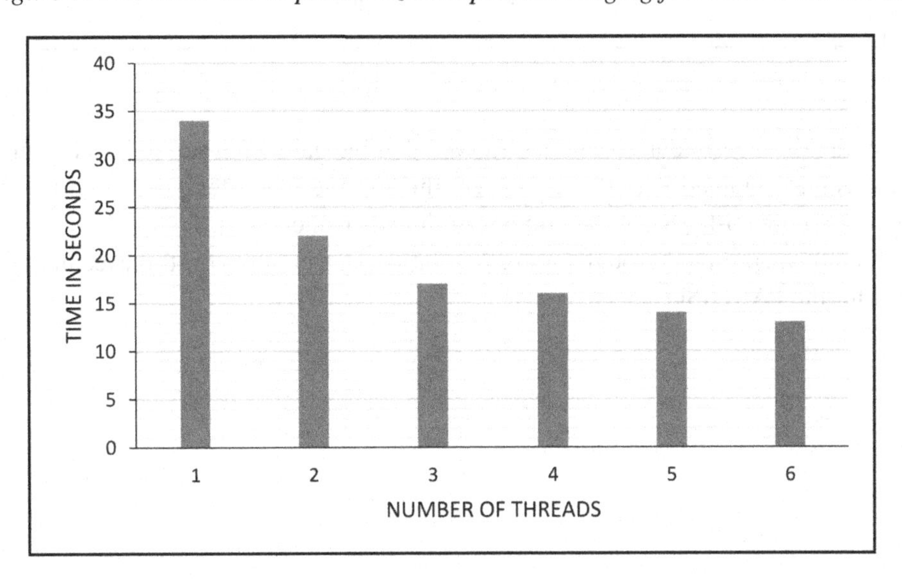

An important time gain can be noted during the execution of PARDISO by using two threads. From this point on, the use of threads increases until we equate the equivalence of one thread for each core. It is also observed that the biggest gain occurs when the solution stops being serial and starts using at least two threads. Figure 6 demonstrates the use of Hyper Threading by simulating the number of threads varying between 7 and 12 threads, the maximum thread capacity available according to the manufacturer.

Figure 6. Solver simulation with hyper threading ranging from seven to twelve threads

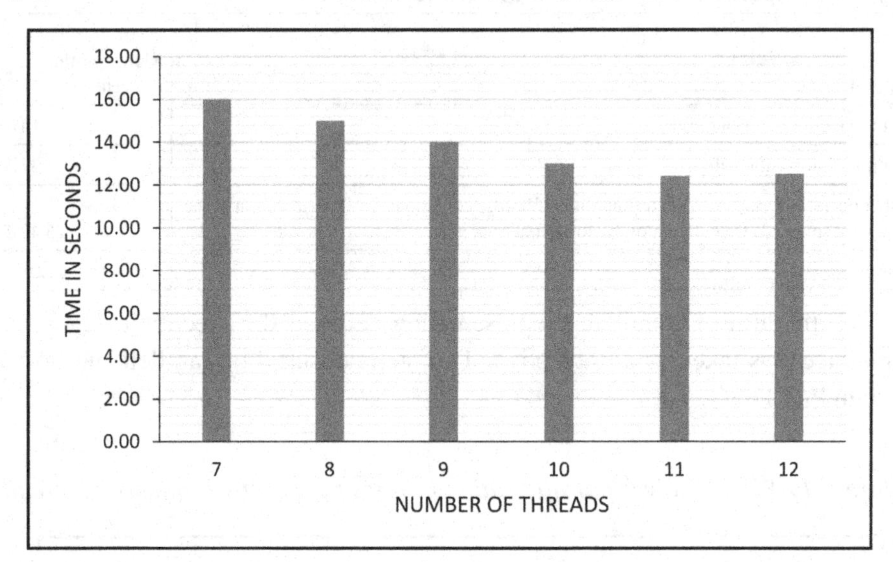

With this experiment, it can be observed that the gain tends not to be as significant as that obtained with the use of real threads and that after a certain point, it becomes more costly when the number of threads is increased.

Finally, we present, through Figure 7, a comparison between the three techniques used for the PARDISO Solver.

Figure 7. Comparison between the parallelization process from one to 16 threads

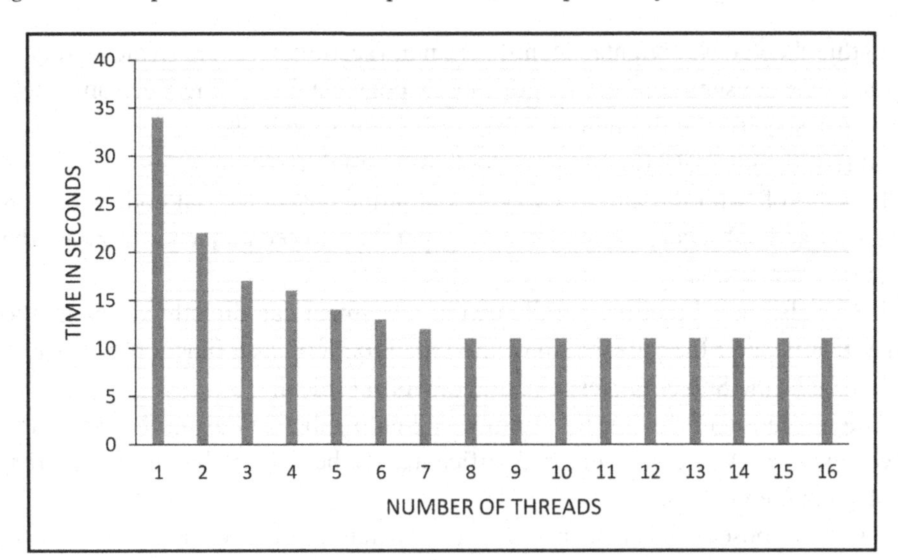

After applying the parallelization techniques presented, we reached the result expressed in Table 4.

Table 5. Result of distribution of the percentage of execution after applied parallelization techniques

Region	Time Consumption Percentage
Initial Settings and Data Entry	15.13%
Loop for integration cells	17.45%
Other operations such as: 1) determination of border conditions; 2) Determination of the support domain; 3) Calculation of the form function; and so on.	44.02%
Solver	23.40%

It can be seen in Table 5 a greater uniformity in time consumption between regions of the code.

CONCLUSION

In this chapter, we sought to identify the main sections with the greatest processing and time consumption in an algorithm responsible for solving electromagnetic problems, specifically through meshless or meshless methods.

During the identification process, tools were used to measure this information, such as GProf, with the aim of carrying out a diagnosis of serial code time consumption and subsequently implementing parallelism techniques in regions that consumed more time.

It was also noted that the parallelism of a segment can directly affect another stage, that is, after the parallelism of the repetition loop as a function of the integration cells, the Solver process became the most critical.

The resolution of the linear system was then parallelized using the PARDISO solver via OpenMP, contributing significantly to the final reduction in the time involved in the process.

It was also observed that the thread limitation must be observed according to each hardware since the elevation of virtual threads can, in turn, increase the execution time, thus breaking the main objective of parallelism which consists of reducing the processing time.

After all processes were executed in parallel, a gain or speedup of 2.4 was noted, representing a significant reduction in the program's time consumption.

REFERENCES

Drmač, Z. (2020). Numerical methods for accurate computation of the eigenvalues of Hermitian matrices and the singular values of general matrices. *SeMA Journal.*, 78(1), 53–92. Advance online publication. 10.1007/s40324-020-00229-8

Ge, G., Wang, X., Manzano, J., & Gao, G. (2009). *Tile Percolation: An OpenMP Tile Aware Parallelization Technique for the Cyclops-64 Multicore Processor*. Springer. .10.1007/978-3-642-03869-3_78

Guimarães, F., Saldanha, R., Mesquita, R., Lowther, D., & Ramirez, J. (2007). A Meshless Method for Electromagnetic Field Computation Based on the Multiquadric Technique. Magnetics. *IEEE Transactions on Magnetics*, 43(4), 1281–1284. 10.1109/TMAG.2007.892396

Idelsohn, S., Oñate, E., Calvo, N., & Del Pin, F. (2003). Meshless finite element method. *International Journal for Numerical Methods in Engineering, 58*, 893-912. .10.1002/nme.798

Li, G. (2005). Meshless Methods for Numerical Solution of Partial Differential Equations. 10.1007/978-1-4020-3286-8_128

Navarro, C., Hitschfeld, N., & Mateu, L. (2013). A Survey on Parallel Computing and its Applications in Data-Parallel Problems Using GPU Architectures. *Communications in Computational Physics*, 15(2), 285–329. 10.4208/cicp.110113.010813a

Nguyen, V. P., Rabczuk, T., Bordas, S., & Duflot, M. (2008). Meshless methods: A review and computer implementation aspects. *Mathematics and Computers in Simulation*, 79(3), 763–813. 10.1016/j.matcom.2008.01.003

Schenk, O., & Gärtner, K. (2002). Two-level dynamic scheduling in PARDISO: Improved scalability on shared memory multiprocessing systems. *Parallel Computing*, 28(2), 187–197. 10.1016/S0167-8191(01)00135-1

Chapter 7
Software Development and Best Practices:
Introduction to Programming Languages Used in Numerical Methods

Ahmed Ibrahim Turki
University of Samarra, Iraq

Sushma Allur
Astute Solutions LLC, USA

Durga Praveen Deevi
Technologies Inc., India

Punitha Palanisamy
https://orcid.org/0000-0003-0727-7072
Tagore Institute of Engineering and Technology, India

ABSTRACT

The growth of science and technology has made it feasible to create software programmes that can quickly solve problems that used to take a long time to resolve. Non-linear equations frequently arise in various complex difficulties that we encounter in our everyday existence. This study aims to assess the effectiveness of the Secant technique and the Newton Raphson method in solving non-linear equations in Python. The programming test was conducted three times with different coefficients and starting values. Another objective of this research is to assess the effectiveness of numerical integration utilising Simpson's methods through the utilisation of Pascal-based software programmes. Two variants of the numerical integration

DOI: 10.4018/979-8-3693-3964-0.ch007

method are the Simpson method. After reviewing research data, the Simpson 1/3 approach proved to be the most accurate strategy to estimate integral values using exponential, polynomial, and trigonometric functions within the 0.001%–0.005% range. The Simpson 1/3 technique runs Pascal faster than Simpson 3/8.

INTRODUCTION

Many models require a theory of equations, especially non-linear equations. Non-linear equations can appear in various fields such as physics, economics, and biology. The nonlinear function most often encountered in applied mathematics is an equation in the form $f(x) = 0$. The function $f(x)$ can be in the form of an algebraic function, transcendent function, or mixed function (Soomro, et al, 2023).

Simple non-linear equations such as the quadratic equation $ax^2 + bx + c = 0$ can be solved in three ways, namely factoring, completing the square, and the formula $abcx_{1,2} = \frac{-b \pm \sqrt{b^2 - 4ac}}{2a}$ Complex non-linear equations cannot be solved analytically usually because it requires a complicated and long process. Therefore, complex non-linear equations require numerical approaches such as Newton Raphson's (NR) method and Secant's method. (Azure, Isaac, et al, 2019) Non-linear equations differ from linear equations in that they involve variables raised to a power other than one, or variable products, which result in curves rather than straight lines when graphed. These equations frequently result in more complex solution methods and can have numerous, zero, or infinite solutions, as opposed to linear equations, which have a single unique solution (if consistent).

The easiest strategies to converge upon are these two. More research was done on the relative effectiveness of these two approaches in terms of execution time, mistakes, and iterations. Along with the rapid development of science and technology, especially in the field of technology, it provides software and hardware devices that are needed today. Engineering results from software and hardware help solve problems quickly. Pascal programming is a software tool that is a creative act in solving problems with a fast-processing process. The development of science and technology cannot be separated from the development of various fields of study, one of which is the field of mathematics (Mahmudova, et al, 2018). Science and technology have evolved to produce programming languages such as Pascal, which have a substantial impact on engineering problem-solving. Pascal noted for its organized approach and ease of understanding, provides for the effective implementation of numerical methods and algorithms, improving problem-solving ability and computational efficiency in engineering activities.

As technology develops, we now provide software tools that can help solve problems working on numerical methods more quickly. Usually, these numerical calculations use the help of applications such as *Matlab, Pascal, and Java* because the calculations are long (Kim, et al., 2003). For example, research from (Papazafeiropoulos, et al., 2016) conducted research on the NR method using MATLAB, (Liu, Zhangjie, et al., 2020) compared the level of convergence of the fixed-point method with the NR method using Matlab. Many help applications besides Matlab can be used for numerical methods. One of them is Python. Python is a programming language that is relatively new in the field of education and is a programming language that is popular in the programming world. This popularity is because Python has code algorithms that are easier to understand and shorter than other programming languages. By using Python for numerical calculations, the program code used can be shorter and faster.

The numerical methods most often used are the NR method and the Secant method because these two methods converge most easily compared to other numerical methods. In previous research, no one has compared the efficiency of these two methods. To determine whether the method is more efficient, investigators used Python to compare the NR and Secant methods in terms of iterations, errors, and program execution time.

According to (Darmawan, R. N. et al., 2016) numerical methods are techniques for solving mathematical problems whose true solutions are difficult to determine analytically. Numerical methods are used to solve mathematical problems regarding integrals called numerical integration. Numerical integration consists of the trapezoidal method, Simpson's method (rule), and Gauss quadratus method. To implement numerical integration calculations in Pascal, define the integral function, set integration limits, choose appropriate numerical methods (such as Simpson's rules), and calculate the integral value using iterative loops and precise arithmetic operations to ensure accuracy. The process involves preparing, creating code, testing, integrating, analyzing, and making results. Numerical integration using the Simpson method has two types of rules, namely $SM\frac{1}{3}$ and $SM\frac{3}{8}$. The $SM\frac{1}{3}$ method (rule) uses a second-order polynomial to approach $f(x)$ while the $SM\frac{3}{8}$ rule uses a +third order polynomial that passes through four points (Vulandari, R. T. et al., 2017).

The Simpson's rule method is widely used by shipping architecture to calculate ship capacity. Apart from that, (Perbani, N. M. R. R. C., et al, 2018) said that the Simpson method is also used to engineer estimates of the area of steam engine indicator diagrams, and surveyors to estimate the area of land plots. The journal only explains the use of the first rule of the Simpson method, namely the $SM\frac{1}{3}$ method, to calculate land area volume and topography. Manual numerical integration, particularly for land area, volume, and topographic computations, is limited due to the enormous time and effort involved. Manual approaches are prone to human error and inefficiency, whereas automated numerical integration utilizing software guarantees improved accuracy, speed, and the capacity to handle complex integrals successfully.

In several cases, there were mathematical calculation problems that in their application were difficult to calculate analytically. Due to their complexity, analytical solutions to sophisticated nonlinear equations are frequently difficult or impossible to obtain. These equations can exhibit features such as many roots, significant sensitivity to initial conditions, and chaotic behavior, making it difficult to find closed-form solutions. The Newton-Raphson and Secant techniques are common numerical approaches for approximating problems. The Newton-Raphson approach often achieves faster convergence due to its quadratic convergence rate, requiring fewer iterations than the Secant method, which has a linear convergence rate. However, the Newton-Raphson technique requires the computation of derivatives, which can be computationally expensive, whereas the Secant method uses finite differences, which is simpler but sometimes slower. Breaking the problem down into smaller, manageable parts, using sophisticated algorithms and computational tools, continuously learning and applying new techniques, and maintaining a positive mindset to stay motivated in tackling challenging problems are some strategies to overcome the complexity of numerical integration problems. Solving numerical integration problems using manual calculations requires quite a long processing time and often this solution gives incorrect final results. Because there are several questions on numerical integration that have a fairly difficult problem level, the manual work process is quite complicated, resulting in a lack of enthusiasm for some people to work on these questions. Therefore, it is easier to use a computer to get answers with the desired accuracy. To apply numerical methods to both integration problems and other problems, technology such as computer programming can be used. In addition, (Herfina, N., et al., 2019) used Pascal programming to calculate the effectiveness of the trapezoid method and Simpson method in determining the area.

discussed the effectiveness of the trapezoid and Simpson method in determining area using Pascal programming. The study focused on comparing the two methods. However, this research will compare the efficiency of the *Simpson* $\frac{1}{3}$ $\left(SM\frac{1}{3}\right)$ and *Simpson* $\frac{3}{8}$ $\left(SM\frac{3}{8}\right)$ methods by paying attention to the efficiency, not only the errors but also the speed of the algorithm when solving problems. used the trapezoid method and $SM\frac{1}{3}$ method and only used polynomial functions Meanwhile, this research uses the $SM\frac{1}{3}$ and $SM\frac{3}{8}$ methods and uses exponential functions, trigonometric functions, and polynomial functions. In this research, we will obtain the efficiency of the $SM\frac{1}{3}$ and $SM\frac{3}{8}$ methods on three types of functions, namely exponential functions, trigonometric functions, and polynomial functions. The efficiency referred to in this research is the speed in solving problems and its accuracy. Implementing iterative algorithms with well-defined convergence criteria, optimizing initial guesses, and employing efficient data structures and computational techniques to minimize errors and runtime ensures accuracy and efficiency in determining equation roots through numerical methods in program code. This research aims to generate a Python program for the NR and Secant methods for non-linear solutions, compare the efficiency of the two methods in solving numerical problems involving polynomial, exponential, and trigonometric functions, and find out how well the current SM methods handle numerical integrals. $SM\frac{1}{3}$ and $SM\frac{3}{8}$ methods with exponential functions, trigonometric functions, and polynomial functions as suggestions for applying the knowledge that has been obtained and to increase insight into the efficiency of numerical integral methods. The steps in the Python program structure for evaluating functions with NR and Secant methods include function definition, initial parameter setting, iterative algorithm implementation, and result output. It is hoped that this research will become reference material for courses in the field of numerical methods and Pascal programming.

METHODOLOGY

To find the roots of non-linear equations, this work uses applied research methodologies, more precisely the NR approach and the Secant method implemented in the Python computer language. The number of iterations, mistakes, and program execution time are the main topics of study. Polynomial, exponential, and trigonometric functions are among the functions that are analyzed.

This research uses pure experimental research (true experimental). The research algorithm used to determine the approximate value of the integral function is:

1. Determine the function to be integrated $y = f(x)$

    1. Enter the lower limit (a) of integration
    2. Enter the upper limit (b) of integration
    3. Determine the number of pias (n) for the $SM\frac{1}{2}$ method which is used (n) even and for the $SM\frac{3}{8}$ method used (n) which is a multiple of 3
    4. Calculate the width of the pias with $h = (b - a)/n$
    5. Determine the exact value of the integral function
    6. Make a table of rules for the $SM\frac{1}{2}$ method
    7. Determine the integration value of the numerical method for the $SM\frac{1}{2}$ method using the formula $I \approx \frac{h}{3}\left(f_0 + 4\sum_{i=odd}^{n-1} f_i + 2\sum_{i=odd}^{n-2} f_i + f_n\right.$ (2.1)

For the $SM\frac{3}{8}$ method, use the formula

$$I \approx \frac{3h}{3}\left(f_0 + 3\sum_{\substack{i=1 \\ i\neq 3,6,9}}^{n-1} f_i + 2\sum_{i=3,6,9}^{n-3} f_i + f_n\right) \quad (2.2)$$

8. Calculate the errors of the $SM\frac{1}{2}$ and $SM\frac{3}{8}$ methods
9. Calculating the processing time for Pascal programming in solving problems.

In this research, there are several steps used to achieve the research objectives, including:

1. Preparation
2. Creating Pascal code based on the $SM\frac{1}{2}$ and $SM\frac{3}{8}$ methods.
3. Try out the $SM\frac{1}{2}$ and $SM\frac{3}{8}$ method programs using several functions for which the answers are known.
4. Revise the program code.
5. Integrate program code into exponential, polynomial, and trigonometric functions to provide estimated results, errors, and processing times.
6. Analyze the results obtained in the program.
7. Conclude.

NR Method Algorithm

The NR method is an iterative method for solving the equation $f(x) = 0$ by assuming f has a continuous derivative f' (Subarinah, 2022:29). The NR method algorithm using the Python programming language is as follows.

i. Determine the function $f(x)$ whose roots will be found.
ii. Determine the maximum iteration epsilon (ε) as the error tolerance.
iii. Determine the value of x_{n-1} as an initial guess
iv. Calculate $f(x_{n-1})$ and $f'(x_{n-1})$.
v. Calculate x_n by $x_n = x_{n-1} - \frac{f(x_{n-1})}{f'(x_{n-1})}$; $f'\left(x_{n-1}\right) \neq 0$, n = 1,2,3,...
vi. If the value $|x_n - x_{n-1}| \leq \varepsilon$ then write $x_{close} = x_n$, otherwise go to iteration $n = n + 1$ and return to step 4.

Secant Method Algorithm

The Secant method is an improvement on the NR method which uses derivatives to find approximations of the roots (Murakami, et al, 2010). The algorithm for the Secant method using Python is as follows.

i. Determine the function $f(x)$ whose roots of the equation will be found.
ii. Determine the maximum iteration epsilon (ε) as the error tolerance.
iii. Determine the values x_0 and x_1 as initial guesses.
iv. Calculate $f(x_0)$ and $f(x_1)$.
v. Calculate x_{n+1} by $x_{n+1} = x_n - f(x_n)\frac{x_n - x_{n-1}}{f(x_n) - f(x_{n-1})} = 1,2,3,\ldots$
vi. If the value $|x_n - x_{n-1}| \leq \varepsilon$ then write $x_{close} = x_n$, otherwise go to iteration $n = n + 1$ and return to step 4.

RESULTS AND DISCUSSION

Solved Equations

Polynomial, exponential, and trigonometric functions are the functions that were used in this study. Based on the function type's behavior and the method's rate of convergence, iteration values are selected. Since Secant's convergence is linear, it takes more iterations than Newton-Raphson's, which is quadratic. According to the properties of each function, certain values are chosen to strike a compromise between computing efficiency and accuracy. The number of iterations and rate of convergence is affected by the specific iteration values selected, which are determined by the behavior of the function and the required precision. By enabling faster computations, higher precision, and the ability to handle more complex integrals, technological advancements, and increased computer power will improve the efficiency of numerical integration techniques. Reducing computing time even more through improved technology and parallel processing will enable real-time integration. The validity of the application software, written in Python and tested using the NR and Secant methods, was evaluated using several functions whose known roots are given in Table 1. The following are the functions in question.

Table 1. Program trial

		Newton Raphson Method			Secant		
Function	**Exact**	**Error**	**Result**	**Iteration**	**Error**	**Result**	**Iteration**
$f(x) = x^2 + 3x + 2$	$x = -1, -2$	-1	0	5	-1	0	7
$f(x) = x^3 - 3x^2 - 6x + 8$	$x = 1, -2, 4$	-2	0	6	-2	0	6
$f(x) = x - e^{-x}$	0.6238	0.6238	0	4	0.6238	0	5
$f(x) = ex - 4x$	0.3931	0.3931	0	4	0.393	0	5
$f(x) = sin(x) + cos(x) + 1$	-1.727	-1.7278	0	6	-1.727	0	5
$f(x) = 2\ sin(x) - cos(x) + 1$	4.4757	4.4757	0	5	4.4757	0	4

The results obtained show that the results of the program are almost 99% the same as the exact values. Therefore, the obtained program is considered valid. Next, to determine the efficiency comparison between the NR method and the Secant method, by raising each function's coefficients and starting values, a program test was run.

Polynomial Function Program

This program is used to determine the roots of polynomial function equations using the NR method and the Secant system in one Python program. The program code used consists of several declarations including the input declaration where the user can input the coefficient values and initial values for which he wants to find the roots of the equation in the format $f(x) = a_1 x^{b_1} + a_2 x^{b_2} + a_3 x^{b_4} + a_4$, the derivative declaration for Newton's method and the declaration for determining roots of equations using NR's method and Secant's method. The NR technique uses explicit derivative calculations and usage, but the Secant method approximates derivatives using finite differences, which simplifies the input requirements. This is the main distinction between the input and derivative declarations in computer code. Using derivative information to facilitate fast convergence, the NR approach iteratively refines guesses to quickly converge on accurate solutions, which is crucial for identifying the roots of polynomial function equations within the Polynomial Function Program. The convergence rate, accuracy (as determined by relative error), computing efficiency, and resilience to changing starting predictions are the performance criteria for both the NR and Secant algorithms. The resulting application program uses an algorithm as in in the following figure. The Newton-Raphson approach is represented in program code via function and derivative declarations, as well as iterative stages based on derivatives. The Secant technique, on the other hand, employs two initial guesses and finite difference approximations, removing the requirement for explicit derivative declarations, which simplifies implementation but may result in more repeats. In the decision section to carry out iteration, the source code is obtained as follows.

Figure 1.

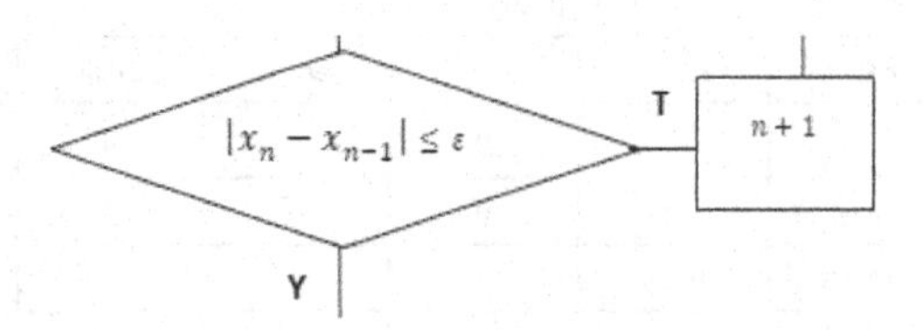

After carrying out the validity test, the root of the function is then searched using different coefficients a_i and b_i. The following are the test results of the polynomial function program using the NR method and the Secant method using Python.

Table 2. Polynomial functions using Newton's Method

Iteration	Error	Root	Time (Seconds)
6	0.00E+00	-1.3318	0.0041
6	0.00E+00	-1.3984	0.0034
5	0.00E+00	-1.6236	0.0031
6	1.00E-05	-1.3692	0.0034
7	0.00E+00	-1.5212	0.0033

Table 3. Polynomial functions using the Secant Method

Iteration	Error	Root	Time (Seconds)
8	0.00E+00	-1.3318	0.0035
7	0.00E+00	-1.3984	0.0042
6	1.00E-05	-1.6236	0.0037
7	0.00E+00	-1.3692	0.0044
7	0.00E+00	-1.5212	0.004

Exponential Function Program

The difference between each function program is the formula or format used. In the exponential function, the format used is $f(x) = a_1 x - a_2 e^{b_1 x}$ using the value a_1 and b_1 = (integer). Considerations for user interface design include usability, clarity, intuitive layout, and accessibility. A well-designed interface improves usability and user experience by offering clear input fields, useful error warnings, and interactive elements to quickly lead users through the computational process. So, in the Python program the only things that need to be changed are the user input and function parts, namely:

$function = a\_one * x + a\_two * sp.exp(b\_one * x)$

The following are the test results of the exponential function program using the NR method and the Secant method using Python.

Table 4. Exponential function using Newton's Method

Iteration	Error	Root	Time (Seconds)
3	0	-0.2059	0.002
3	0	0.361	0.0012
2	0	-0.3552	0.0032
6	0	-0.4306	0.0029
6	0	-0.4306	0.0042

Table 5. Exponential function using the Secant Method

Iteration	Error	Root	Time (Seconds)
4	1.00E-05	-1.6024	0.0022
3	0.00E+00	0	0.0051
3	0.00E+00	-0.6232	0.0044
7	1.00E-05	3.6074	0.0032
10	0.00E+00	3.2933	0.0021

Trigonometric Function Program

The format for trigonometric functions is $f(x) = a_1 sin(b_1 x) + a_2 cos(b_2 x) + a_3$ using the value a_1 and b_1 = (integer). The only Python program that needs to be changed is the user input part and its function, namely:

$function = a\_one * sp.sin(b\_one * x) + a\_two * sp.cos(b\_two * x) + a\_three$

The following are the test results of the trigonometric function program using the NR method and the Secant method using Python.

Table 6. Trigonometric functions using Newton's Method

Iteration	Error	Root	Time (Seconds)
4	0	-1.5865	0.0022
3	0	0	0.005
3	0	-0.617	0.0044
4	1.00E-05	3.5717	0.0032
1	0	3.2607	0.0021

Table 7. Trigonometric functions using the Secant Method

Iteration	Error	Root	Time (Seconds)
3	0	-1.5865	0.0049
5	0	0	0.0051
5	0	-0.617	0.0039
6	0	3.5717	0.0048
4	0	3.2607	0.0035

Implications of Results

Based on several experiments above, determining the root of the function in this study uses the NR and the SM methods. The results of this experiment are that the Newton method has a shorter number of repetitions, mistakes, and implementation time than the Secant method. The findings for Functions in the Secant Method discussion section demonstrated adequate performance in terms of accuracy and speed of convergence, confirming the method's suitability for these specific instances even in the absence of derivative computations. Because of its quadratic convergence rate, the Newton approach typically takes less iteration than the Secant method. Each iteration of the Newton technique significantly reduces the error, resulting in faster convergence when the starting approximation is somewhat close to the true root. Specifically, each function type's iteration values strike a balance between computing efficiency and accuracy. The order of the functions used also affects the number of iterations in each method. This is in line with the opinion of regarding the increase in function order. When utilizing numerical methods to converge, higher-order functions typically require more iterations. The convergence rate is influenced by the function's complexity, with higher-order polynomials frequently requiring more iterations to reach the required precision. The larger the function order used, the smaller the error generated so that the results will be more accurate. The explanation of each table is as follows.

1. Table 2 and Table 3 when compared, the iteration of the NR method is shorter than the Secant method. The error for Newton's method is also smaller than Secant's method. The average time required by the Newton method is less than the Secant method. When considering the number of iterations, errors, and average execution time, the Newton technique is more efficient than the Secant approach.
2. Tables 6 and 7 show that for the exponential function, the NR approach has fewer iterations, errors, and execution time than the Secant technique, even though the coefficients and beginning values differ. The Newton method iteration is 22% more efficient and 12% more time efficient than the Secant method. The quadratic convergence rate of the Newton technique reduces the number of iterations required to obtain the same precision, resulting in a 12% increase in time efficiency when compared to the Secant approach. This saves computational time.
3. Program test results for trigonometric functions with different coefficients and initial values show that the NR method iterations are shorter than the Secant method. while for error, there is a difference of 0.00001 in the NR method. The time used by the Newton method is shorter than the Secant method.

The program code used in the Pascal software program was created by following the algorithm in the $SM\frac{1}{3}$ and $SM\frac{3}{8}$ methods. Therefore, a program trial was carried out using integral exponential functions, polynomial functions, and trigonometric functions to see the accuracy of the calculations obtained from the Pascal software program using the solution to a previously known function. Ensuring proper function definitions, adequate starting guesses, and unambiguous input formats are necessary to modify the user input and function parts of the exponential function software to properly manage the unique properties of exponential functions. The trial results of the $SM\frac{1}{3}$and $SM\frac{3}{8}$ methods using the Pascal software program on integral exponential functions, polynomial functions, and trigonometric functions are presented in Table 8 below:

Table 8. Trial of the ***SM*** $\frac{1}{3}$ *and* ***SM***$\frac{3}{8}$ *method program*

		SM 1/3			SM 3/8		
Function	**Exact**	**n**	**Approximate**	**Error (%)**	**n**	**Approximate**	**Error (%)**
$\int_0^4 e^x dx$	55.2	20	55.207	0.001	21	55,20,699	0.0022
$\int_1^5 (2x^2 - 3x + 3)^2 dx$	1354.7	20	1354.66	0.0003	19	1283.9	5.72
$\int_1^2 (4 + cos\, x)^2 dx$	41.3	10	41.288	0.0001	9	41.3	0.0003

Pascal programming used to calculate the numerical integral efficiency of the $SM\frac{1}{3}$and $SM\frac{3}{8}$ methods displays output in the form of commands to enter the upper limit (a) and lower limit (b) of the integral function to be calculated, then enter the iteration (n) which will be calculated with each segment having an even value for the $SM\frac{1}{3}$ method and multiples of 3 segments for the $SM\frac{3}{8}$ method. The main goal of using time measurement while the SM 1/3 and SM 3/8 approaches execute basic algorithms is to assess their computational efficiency and make sure they produce accurate results in reasonable amounts of time, which is essential for real-world applications. Next, the user presses the enter button and the output will come out in the form of iterations n, exact, approximate value, relative error, and processing time. By using numerical techniques to test functions, recording the outcomes, and comparing them to known solutions, the software program estimates values, relative mistakes, and processing times.

Numerical integration of the $SM\frac{1}{3}$ and $SM\frac{3}{8}$ methods on integral exponential functions implemented in Pascal-based software programs can retrieve estimated values, relative errors, and processing durations by entering varying iterations (n) and exact values. The upper limit (a) and lower limit (b) are used is $\int_0^{10} f(x)dx$ *with* $f\left(x\right) = \frac{a \times e^{bx}}{c \times e^{dx} + e_1}$ *where* $a, b, c, d, e_1 = \left[1,2,3,4.5\right]$.

The results of the numerical integration calculations can be seen in Table 9 below:

Table 9. Numerical integration of the $SM\frac{1}{3}$ *and* $SM\frac{3}{8}$ *exponential function methods*

		SM 1/3				SM 3/8			
	Exact	n	Approximate	Error (%)	Time (milliseconds)	n	Approximate	Error (%)	Time (milliseconds)
1	0.385	55	0.3773	0.004	3.96	67	0.3773	0.004	5.87
2	0.795	59	0.7794	0.006	4.499	69	0.8488	0.006	6.03
3	16.92	18	16.5911	0.001	2.409	23	18.0695	0.001	3.15
4	47.93	18	46.9997	0.002	2.541	30	51.1877	0.001	3.42
5	0.392	79	0.3844	0.004	6.413	100	0.4187	0.004	7.68
6	0.38	46	0.373	0.006	2.794	63	0.4063	0.007	4.9
7	2.404	66	2.3575	0.006	5.83	78	2.5675	0.007	7.08
8	1.187	46	1.1641	0.004	2.75	63	1.2678	0.006	4.72
9	1.061	33	1.0404	0.003	3.487	39	1.1331	0.006	3.81
10	0.121	90	0.1189	0.006	6.435	98	0.1295	0.007	7.41

Apart from integrals of exponential functions, Pascal programming can be implemented on integrals of polynomial functions. The upper limit (a) and lower limit (b) used are $\int_0^{10} f(x)dx$ with $f(x) = a \times x^5 - b \times x^4 + c \times x^3 + d \times x^2 - e_1$. *where* a, b, c, d, e_1 $\left[1,2,3,4.5\right]$. The results obtained from the Pascal programming calculation process for integral polynomial functions can be seen in Table 10 below:

Table 10. Numerical integration of ***SM*** $\frac{1}{3}$ *and* ***SM*** $\frac{3}{8}$ *polynomial function methods*

	SM 1/3				SM 3/8			
Exact	**n**	**Approximate**	**Error (%)**	**Time (milliseconds)**	**n**	**Approximate**	**Error (%)**	**Time (milliseconds)**
779164	22	764045	0.001	3.41	30	764042	0.001	4.24
784335	20	769120	0.002	3.3	23	837660	0.002	3.36
139514	24	136806	0.001	3.6	30	148996	0.001	4.64
271910	22	266634	0.001	3.47	23	290397	0.003	3.56
334729	26	328232	0.001	3.94	28	338279	5.908	4.27
789485	24	774162	0.001	3.55	32	824839	2.387	4.93
830674	15	814588	0.006	2.52	17	887208	0.01	2.56
482534	20	473174	0.002	2.94	23	515340	0.002	3.3
315685	15	309572	0.006	2.56	17	337170	0.01	3.22
477395	22	468131	0.001	3.52	23	509851	0.002	3.66

Apart from using integral exponential functions and polynomial functions, integral trigonometric functions can also be implemented in the Pascal software program. The selection of particular integral functions was justified by their capacity to illustrate the effectiveness and precision of the integration techniques. A variety of functions with different levels of complexity are used to verify the dependability and adaptability of the numerical methods used. The integral of the trigonometric function used has an upper limit (*a*) and a lower limit (*b*), namely $\int_0^{\pi} f(x)\, dx$ *with* $f(x)\ a \times sin\ (bx) + c + cos\ (dx) + e_1$ *with* $a.b.c.d.e_1 = [1 - 5]$. The output produced by the Pascal software program for integral trigonometric functions can be seen in Table 11 below:

Table 11. Numerical integration method of ***SM*** $\frac{1}{3}$ *and* ***SM*** $\frac{3}{8}$ *trigonometric functions*

	SM 1/3				SM 3/8			
Exact	**n**	**Approximate**	**Error (%)**	**Time (milliseconds)**	**n**	**Approximate**	**Error (%)**	**Time (milliseconds)**
3.445	11	3.445	0.004	2.64	17	3.445	0.002	3.443
7.633	24	7.633	0.002	3.102	30	7.633	0.002	3.223
11.249	31	11.249	0.004	3.377	36	11.25	0.006	3.399
14.26	26	14.26	0.003	3.113	30	14.26	0.006	3.344
18.583	33	18.584	0.003	3.553	36	18.584	0.006	3.795
6.91	11	6.91	0.001	2.585	17	6.91	0.001	3.443
14.753	13	14.753	0.001	3.025	17	14.753	0.001	3.443
14.025	37	14.025	0.001	3.729	43	14.025	0.001	3.982
6.393	33	6.393	0.001	3.916	63	6.393	0.001	5.016
4.77	46	4.771	0.003	3.872	56	4.771	0.003	5.225

Based on Tables 8, 9, and 10, it can be seen that each iteration (n) and exact have different approximate values, but several of the n iterations of the exponential function, polynomial function, and trigonometric function have the same relative error. The tolerance value given to determine the iteration (n) of the function is efficient if the relative error obtained in each calculation is less than or equal to 5×10^{-5}. It is possible to obtain a relative error of less than 5×10^-5 by increasing the number of iterations, improving first estimations, utilizing higher precision arithmetic, and utilizing adaptive algorithms that dynamically modify step sizes in response to error estimates. Some of these iterations have a relative error of less than 5×10^{-5} or less than 0.005% so it can be assumed that the iteration (n) is efficient. Numerical instability, behavior of the function close to critical points, or insufficient iterations can all contribute to a relative inaccuracy in integral functions larger than 0.005%. To lessen this problem, make sure the algorithms are reliable and have enough iterations. Meanwhile, integral functions that have a relative error greater than 0.005% can be said to be inefficient because the approximate value is still far from the actual value so higher iterations (n) are needed to produce a relative error of less than 5×10^{-5} or less than 0.005%.

Discussion of numerical integration calculations using the $SM\frac{1}{3}$ and $SM\frac{3}{8}$ to obtain approximate values, namely values that are close to the true value of the integral of exponential, polynomial, and trigonometric functions by implementing these calculations into Pascal-based software programs.

In this study, a second-order polynomial was used to approximate $f(x)$ for the $SM\frac{1}{3}$ method. The value of the $SM\frac{1}{3}$ method is obtained because Δx is divided into 3, which will produce an approximate value that is close to the actual value and produces a different error. According to (Murakami, et al, 2010), the $SM\frac{1}{3}$ method uses second-order polynomials to approach the curve $f(x)$ which passes through the point $(x_{i-1}, f(x_{i-1}))$ $(x_i, f(x_i))$ $(x_{i+1}, f(x_{i+1}))$. Therefore, in this study, it was used. Iteration (n) has an even value to get an approximate value for each problem. Meanwhile, the $SM\frac{3}{8}$ method uses a third-order polynomial by going through four points to get an approximate value that is close to the true value and produces different errors.

Based on the research results, a program trial has been carried out using the $SM\frac{1}{3}$ and $SM\frac{3}{8}$ methods following the research algorithm by entering integral exponential functions, polynomial functions, and trigonometric functions into the program. The numerical outputs are compared with high-precision benchmarks or known exact values to determine how accurate the computations were in the program trials. To measure the departure from perfect solutions and guarantee the program's dependability, relative errors and residuals are calculated.

Pascal-based software is a problem that will be tested to find out if the approximation and error values obtained from the Pascal software program calculations are the same as the approximation and error values obtained from calculations that have been carried out previously. Pascal-based software has several benefits, such as an organized approach, ease of learning, and the capacity to manage intricate numerical calculations with efficiency, all of which improve mathematical learning and problem-solving abilities. Therefore, the source code used can be said to be valid and can be used for exponential functions, polynomial functions, and other trigonometric functions. In line with the results of research conducted by the Pascal software program can be used to calculate the numerical integration of the Simpson method. In addition to running programs on different functions, theoretical analysis of error bounds and convergence properties, as well as comparison with other numerical integration techniques such as the trapezoidal or Gauss quadrature methods, might be employed to ascertain the efficiency of the Simpson method. Using this software can increase learning motivation, numerical abilities, and problem-solving abilities for mathematical problems.

It is possible to determine which Simpson method is more efficient for calculating the integral of a function by running programs on various functions, such as exponential, polynomial, and other trigonometric functions, and comparing the results in terms of processing time, relative errors, and approximate values. Algorithm optimization, effective coding techniques, and the use of computational resources are all necessary to reduce processing time while preserving accuracy. The polynomial, exponential, and trigonometric functions are among those utilized to assess the correctness of the application program. To evaluate the efficiency of the approaches and make sure they work well in real-world scenarios, processing time is evaluated during the execution of the core algorithm. The relative error threshold determines iteration efficiency and guarantees correctness within reasonable bounds.

The time obtained during the execution process carried out by the Pascal software program to obtain approximate values for exponential, polynomial, and trigonometric functions is different for each iteration (n) used. This is because the higher the number of iterations (n) used in the integral function, the longer the time needed to solve the problem. Therefore, the higher the iteration (n) given to each problem of integrating exponential functions, polynomial functions, and trigonometric functions, the longer it will take Pascal programming to solve the problem and the approximate value obtained will be closer to the true value and error. the output will be smaller. The iteration (n) used for each exponential function integral, polynomial function, and trigonometric function to find the numerical integration for each exponential function integral will be said to be efficient if the approximation value obtained is close to the actual value and the relative error obtained at each iteration is no more than the tolerance limit. or predetermined stopping criteria, by research conducted by Perbani & Rinaldy (2018) that requires consideration of the iteration size (n) or segments to determine the efficiency of the calculation process using the $SM\frac{1}{3}$ method. The link between iteration size and processing time in numerical integration algorithms is influenced by a number of factors that should be looked at, including hardware performance, function complexity, convergence rate, initial guess quality, and algorithm type. Research carried out to obtain approximate values, errors relative and processing time on integral functions

Exponential, polynomial functions, and trigonometric functions are the opinions of (Ermawati, E., et al, 2017) who say that the greater the use of iteration (n), the better the results given tend to be. Therefore, the iteration (n) used to calculate the approximation value for each integral function is very high so that an approximation value is obtained that is close to the actual value and a smaller error is obtained.

In this study, using the $SM\frac{3}{8}$ method obtained greater approximation and error values compared to using the $SM\frac{1}{3}$method as well as the processing time required to obtain the approximate value of the function, there is quite a big difference. Therefore, the level of accuracy of the $SM\frac{1}{3}$ method obtained from calculating the results of numerical integrations is more accurate than the $SM\frac{3}{8}$ method. This explanation is by the opinion of (Ponalagusamy R, et al., 2010) who said that the $SM\frac{1}{3}$rule is preferable compared to the $SM\frac{3}{8}$ rule because it reaches a level of accuracy up to order three with just three points. More the iteration (n) used, the smaller the error produced and the more accurate the results will be. In line with Ermawati's opinion that the greater the use of (n) iterations, the better the results tend to be.

Time measurement is only used when executing the core algorithm using the $SM\frac{1}{3}$ and $SM\frac{3}{8}$ methods. The algorithm used is the algorithm when determining approximate values for exponential functions, polynomial functions, and trigonometric functions using the $SM\frac{1}{3}$ and $SM\frac{3}{8}$ methods. The time required during the process of working on exponential functions, polynomial functions, and functions

The trigonometry in each iteration (n) is not too different, there is even the same processing time duration in several iterations (n) in the integral function. In line with research conducted by Herfina (2019) the accuracy of algorithm execution time is obtained by not calculating the time required to display income or expenditure operations in reading or writing, the start of the program, and so on. The unit of time used in this algorithm is milliseconds.

Based on the results of calculations using the $SM\frac{1}{3}$and $SM\frac{3}{8}$ methods, it was found that there is an influence on the time required to execute Problems with exponential functions, polynomial functions, and trigonometric functions based on the number of iterations (n) used in the problem. The calculation results obtained from the Pascal software program to find the approximate value of the integral function show that the more iterations (n) used, the greater the time needed to process the algorithm. This is because the more iterations (n) that are included in the Pascal program, the more steps in the algorithm that must be completed and the more time required in the process.

The processing time used in the $SM\frac{1}{3}$ and $SM\frac{3}{8}$ methods has quite a large difference in processing. The resulting difference in execution time. This problem takes around 3-5 milliseconds per attempt to run a Pascal program on each integral function. This is because the $SM\frac{1}{3}$ and $SM\frac{3}{8}$ methods have slightly different work algorithms; the difference lies in the equation formula for the $SM\frac{1}{3}$ and $SM\frac{3}{8}$ methods.

CONCLUSION

Based on the aforementioned facts, it was deduced that the discussion led to the development of a Python-based application program for determining the roots of polynomial, exponential, and trigonometric functions. For polynomial functions, the Newton approach outperforms the Secant technique by 16%. For exponential functions, the Newton technique is 22% more efficient than the Secant method. When it comes to trigonometric functions, the Newton approach outperforms the Secant technique by 24% in terms of iterations. As a result, for these three sorts of functions, Newton's method outperforms Secant's. In this study, polynomial functions, exponential functions, and trigonometric functions are identified as the equation's roots. It is hoped that future studies will cover fourth-degree polynomial functions and so on. It was also found that the $SM1/3$ method was more efficiently used to find approximate values for integrals of exponential, polynomial, and trigonometric functions.

DECLARATIONS

Funding

No funds or grants were received by any of the authors.

Conflict of Interest

There is no conflict of interest among the authors.

Data Availability

All data generated or analyzed during this study are included in the manuscript.

Code Availability

Not applicable.

Author's Contributions

All authors contributed to the design and methodology of this study, the assessment of the outcomes, and the writing of the manuscript.

REFERENCES

Azure, I., Aloliga, G., & Doabil, L. (2019). Comparative Study of Numerical Methods for Solving Non-linear Equations Using Manual Computation. *Mathematics Letters.*, 5(4), 41. 10.11648/j.ml.20190504.11

Darmawan, R. N. (2016). Comparison of the Gauss-Legendre, Gauss-Lobatto, and Gauss-Kronrod Methods for Numerical Integration of Exponential Functions. *Journal of Mathematics and Mathematics Education*, 1(2), 99–108.

Ermawati, E., Rahayu, P., & Zuhairoh, F. (2017). Comparison of Numerical Solutions for Double Integrals in Algebraic Functions using the Romberg Method and Monte Carlo Simulation. *MSA Journal*, 5(1), 46–57.

Herfina, N., Amrullah, A., & Junaidi, J. (2019). The Effectiveness of the Trapezoidal and Simpson Methods in Determining Area Using Pascal Programming. *Mandalika Mathematics and Education Journal*, 1(1), 53–65. 10.29303/jm.v1i1.1242

Kim, A., Park, C., & Park, S.-H. (2003). Development of Web-based Engineering Numerical Software (WENS) Using MATLAB: Applications to linear algebra. *Computer Applications in Engineering Education*, 11(2), 67–74. 10.1002/cae.10038

Liu, Zhang, Su, Sun, Han, & Wang. (2020). Convergence Analysis of Newton-Raphson Method in Feasible Power-Flow for DC Network. *IEEE Transactions on Power Systems*. IEEE. .10.1109/TPWRS.2020.2986706

Mahmudova, S. (2018). *Features of Programming Languages and Algorithm for Calculating the Effectiveness*. Academic Press.

Murakami, A., Wakai, A., & Fujisawa, K. (2010). Numerical Methods. *Soil and Foundation*, 50(6), 877–892. 10.3208/sandf.50.877

Papazafeiropoulos, G. (2016). Newton Raphson Line Search - Program for the solution of equations with the quasi-Newton-Raphson method accelerated by a line search algorithm. 10.13140/RG.2.1.3485.5282

Perbani, N. M. R. R. C., & Rinaldy. (2018). Application of Simpson's Volume Calculation Method to Calculate Ship Volume and Land Topography. *Journal of Green Engineering*, 2(1), 90–100.

Ponalagusamy, Pja, & Muthusamy. (2010). Numerical Methods on Ordinary Differential Equation. *International Conference on Emerging Trends in Mathematics and Computer Applications 2010, 1.*

Soomro, S., Shaikh, A., Qureshi, S., & Ali, B. (2023). A Modified Hybrid Method For Solving Non-Linear Equations With Computational Efficiency. *VFAST Transactions on Mathematics.*, 11(2), 126–137. 10.21015/vtm.v11i2.1620

Vulandari, R. T. (2017). *Numerical Methods: Theory, Cases, and Applications*. Mavendra Pres.

Chapter 8
Solving Linear Problems: Applications in Structural, Thermal, and Fluid Analysis

Alnoman Mundher Tayyeh
Mangalore University, Iraq

Akram H. Shather
Sulaimani Polytechnic University, Iraq

Firas Tarik Jasim
Al-Dour Technical Institute, Northern Technical University, Iraq

Saja Sumiea Anaz
Mangalore University, Iraq

ABSTRACT

Problems associated with heat exchange between a gas and a packed bed are quite common in several branches of industry. To properly study this type of phenomenon, a good prediction of the flow dynamics through the bed is necessary. Based on this, this work presents a study of the flow in a packed bed through computer simulation. The packed bed considered in this study is cylindrical and made up of spheres. The simple cubic sphere arrangement was considered to generate the bed geometry. To make the simulation feasible, some simplifications were necessary. The first was flow periodicity in the axial direction of the bed. Another simplification was symmetry in relation to one-eighth of the bed. A very simplified case, considering a single simple cubic cell of spheres, was also simulated. The geometry and mesh of the packed bed were generated in the Gambit software and the flow simulation was carried out in the Fluent software.

DOI: 10.4018/979-8-3693-3964-0.ch008

INTRODUCTION

Problems arising from a flow crossing a packed bed are common in several branches of industry. Packed bed flow is characterized when a fluid impinges on the particles and is unable to promote their movement, in such a way that these particles can reach the critical fluidization speed. In this case, the fluid only travels through the spaces (pores) existing in the bed. According to Graciano-Uribe (2021), the height or size of the packed bed influences the pressure drop and the flow regime. To carry out a flow assessment and determine the pressure drop, it is necessary to establish the dimensions of the bed and the incident air flow.

Keyser et al. (2006) addresses the role of coal particle size distribution on the pressure drop of a packed bed. The study was carried out using computational fluid dynamics (CFD) and the results obtained in the simulation were compared with the results obtained through the Ergun equation. The comparison between the two methods showed that for distributions with the same average diameter, porosity and sphericity, the pressure drops values present the same result given by the Ergun equation. The structures of the coal beds were simulated assuming that the coal particles were convex polyhedra arranged randomly in three-dimensional space. Therefore, two beds composed of randomly shaped polyhedra were simulated, manipulated to obtain the same sphericity, porosity and particle size distribution, but with a different particle size distribution range. The authors concluded that the bed with a wider range of particle size distribution causes a greater pressure drop, for the same sphericity and porosity. They also concluded that CFD simulation can be used to adjust the empirical constants of the Ergun equation and that it is not suitable for wide ranges of particle size distribution.

With the same objective of analyzing a method to determine the pressure drop across a packed bed, Barton et al. (2013) uses CFD techniques to simulate a porous bed of uniform spherical particles. The work was based on a thermal energy storage system through airflow under a bed of rock. To characterize the porous media, three main parameters were determined: thermal conductivity, particle and fluid heat transfer coefficient, pressure drop and porosity. According to Cascetta, et al (2016), bed assemblies were limited in their capabilities to determine heat transfer parameters in a fixed bed. In this way Cascetta, et al (2016) found an increase in porosity as the particle diameter-cylinder diameter ratio increased, for all arrangements. He also mentioned that the values found through the simulation were very close to the values found through the Eisfeld/Schnitzlein equation for the pressure drop.

Pesic, et al (2015) question the use of general equations for beds of non-spherical particles and non-uniform sizes. A significant portion of the surface area of the particles cannot be reached by the flow because they overlap each other. Another point addressed is the fact that for beds tightly packed in a vertical cylindrical column, the main orientation of the particles is almost horizontal, resembling a line structure. Therefore, the fluid path, for a given thickness, is longer through this type of bed than through a bed of isotropic particles.

BEDS PACKED WITH SPHERES

A packed bed can be defined as a fixed (or nearly fixed) solid matrix with connected voids through which a fluid can flow, constituting a porous medium (Nemec & Levec, 2005).

Figure 1. Schematic of a Packed Bed

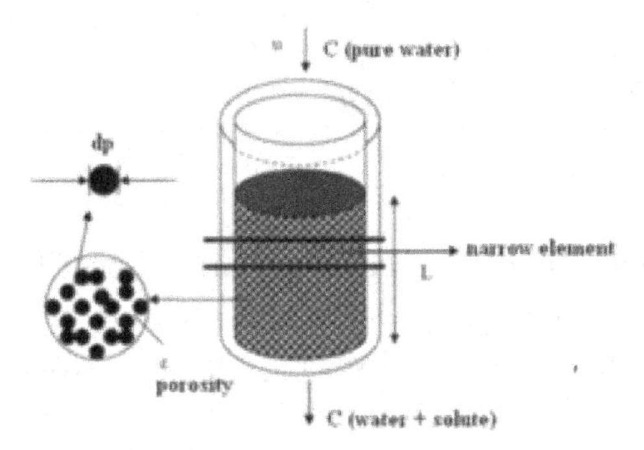

These empty spaces, called pores, have different arrangements, enabling the condition of non-uniformity in the paths in which the fluid travels. Porosity is the percentage of volume not occupied by particles, in relation to the entire volume of the bed. In Figure 2, the fluid flow rate that passes through area A of the bed is Q. The existence of pores within the bed reduces the area for fluid flow. To preserve the continuity of the fluid flow with the flow that enters the surface area of the bed, there will be a narrowing of the streamlines after a reduction in the surface area, thus the velocity (V_m) inside the bed will be greater than the surface, also called the Darcy V_D velocity (Winter et al., 2022).

Figure 2. Schematic about Darcy Speed and Intrinsic Average Speed

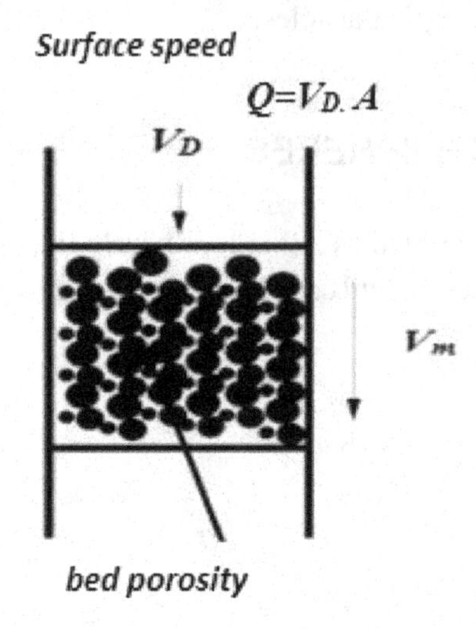

Darcy speed is not a physical speed, it is a surface speed based on the complete cross-section of the medium, not just the cross-section of the fluid. The Darcy velocity and the mean intrinsic fluid velocity, V_m, are related to the bed porosity (ε) as follows.

$$V_D = \varepsilon V_m \tag{1}$$

A parameter of great importance in the evaluation of an internal flow in a porous medium is the pressure drop. When a fluid flows from one point to another inside a tube, there will always be a loss of pressure, called a pressure drop. This pressure loss may arise, for example, from the friction of the fluid with the internal surface of the tube wall. Therefore, the greater the roughness of the pipe wall or the more viscous the fluid, the greater the pressure loss will be (Graciano-Uribe, 2021). The pressure loss will be equal to the pressure drop if:

• • The fluid is incompressible (ρ=constant);

Thus, knowing that V1 and V2 are the velocities at the inlet and outlet respectively and that the volumetric flow at the inlet must be the same as that at the exit of the section:

$$V_1 A_1 = V_2 A_2 \tag{2}$$

Consequently, V_1 is equal to V_2. From the modified Bernoulli Equation, Eq. (3), it is inferred that the energy loss is equal to the pressure drop, Eq. (4):

$$P_1 - P_2 = \rho g(Z_1 - Z_2) + \frac{1}{2}\rho(V_1 - V_2) + h \tag{3}$$

$$P_1 - P_2 = h \tag{4}$$

Packaged Bed of Reference Work

The bed studied by Graciano-Uribe (2021) will serve as a reference for the present study. The bed was 0.15 m in diameter and the length of the packed volume was variable. The study methodology used by Graciano-Uribe (2021) was experimental. Pressure drop data was collected for the analysis of head loss, before the flow impacted the bed and immediately after the flow left it. Several pressure measurements were also carried out for different bed lengths, describing the process according to this variation. overlapping and separated only by steel screens, that is, at each interposition of the bed length a a screen to separate the bed. Starting from the empty tube, reaching a total of seven beds depending on the diameter of the tube at a rate of 0.5 D, as shown in Figure 3, totaling in the end a bed length of 0.525 m.

Figure 3. Porous Bed Under Study

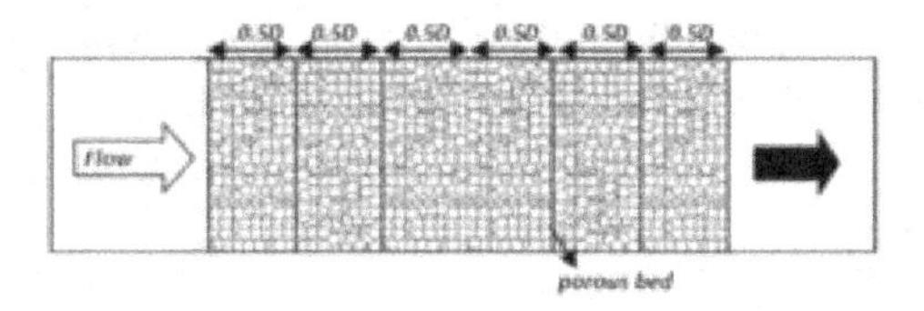

From a sample of 230 pits, Graciano-Uribe et al. (2010) measured the average diameter of the açaí seed, as indicated in Figure 4. The data obtained can be seen in Table 1. After obtaining the average diameter, obtained depending on each direction, an average of the three diameters was taken and the result was a diameter of 1.03 cm, this value being approximately 1.0 cm for the main case of this work.

Figure 4. Directions Adopted for the Average Diameter

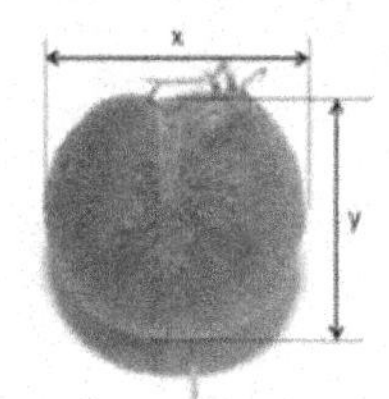

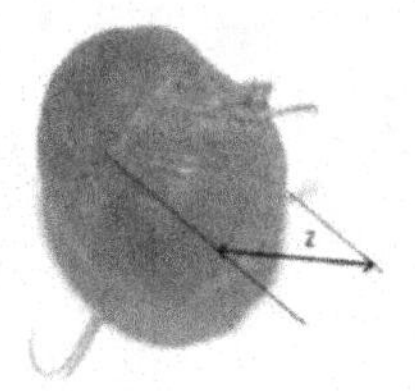

Table 1. Average Diameter Values Adopted

Directions	X	Y	Z
Average diameter (cm)	11,554	0.974	0.965
Standard Deviation (cm)	0.0684	0.0834	0.0644

NUMERICAL SOLUTION METHODOLOGY

In this part of the chapter, the methodology used to construct and simulate the flow through the packed bed of spheres will be presented.

Governing Equations

To analyse the flow, equations that govern the physical problem are used: Equations of Conservation of Mass and Quantity of Motion.

$$\partial\rho/\partial t + \nabla.(\rho\vec{v}) = 0 \quad (5)$$

Eq. (5) is the general form of the mass conservation equation, valid for compressible and incompressible flows. The Mass Conservation Equation, or Continuity Equation, corresponds to the sum of the "rate of mass variation within a control volume" and the "mass flow that crosses the control surface" equal to zero. In the case of incompressible flow, density is not a function of time or space, therefore, $\partial\rho$ $\partial\tau \cong 0$. In the second term on the left side, the p-(constant) of the divergent operator can be removed. Thus, Eq. (5) reduces to:

$$\nabla.(\vec{v}) = 0 \quad (6)$$

The meaning of each term in Eq.(7) is given as follows: the first term is the variation in momentum, which can also be understood as the force per unit volume of an infinitesimal particle since it is the product of density and acceleration material; the second and third terms refer to the surface forces that act on the infinitesimal control volume, they represent the action of the forces resulting from the stress tensor in the fluid which is subdivided into two parts: one resulting from the pressure field and the other due to the deformations given by the velocity field; finally the fourth term represents the gravitational field strength per unit volume.

Simulation Code

In this work, the commercial computational fluid dynamics code Fluent was used to simulate the flow. To calculate the flow field, the code uses a numerical technique based on control volumes, to which the principle of conservation of a given scalar quantity is applied, that is, the code solves the governing flow equations for each finite volume. The discretized equations are obtained by integrating the governing equation over each of the control volumes in the domain.

According to Eymard (2000), the finite volume method consists of:

- Division of the computational domain into a mesh composed of discrete control volumes.
- Integration of the governing equations over each control volume resulting in equations that contain discrete variables such as: speed, pressure, temperature, etc.
- Linearization of discretized equations and solution of the system of linear equations.

Variable values are stored in the center of the control volumes. On the faces of these volumes, the values of discrete variables are expressed using interpolation functions.

Choosing the Type of Packed Bed Arrangement

In the work of Graciano-Uribe (2021) it is difficult to define the arrangement of spheres throughout the packed bed. However, for numerical simulation it is necessary to define an arrangement type for generating the bed geometry and this must be done by minimizing the complexity of all geometric development, thus, three types of arrangements were selected for generating the packed bed, shown in Figure 5. Each arrangement is formed by basic and repetitive units called unit cells, which have their own characteristics:

- Face Centered Cubic (CFC). In the CFC arrangement there is an eighth of a sphere at each vertex and half a sphere at the center of each face of the cube. In each cell there are four spheres and porosity equal to 0.26. In this type of cell, the edge L of the cube has $2\sqrt{2}$ times the radius R of the sphere.
- Primitive cubic (CS). In this type of packing there is one-eighth of each sphere at each vertex of the cube, that is, the cell allocates one atom. The edge L of the cube has twice the radius R of the sphere and porosity equal to 0.48.
- Body Centered Cubic (CCC). The CCC arrangement has a 1/8 sphere at each vertex and a sphere at the center of the cube. In each cell there are two spheres. In this type of cell, the edge L of the cube has $4/\sqrt{3}$ times the radius R of the sphere and porosity equal to 0.32.

Figure 5. CS, CCC, CFC Geometries Respectively

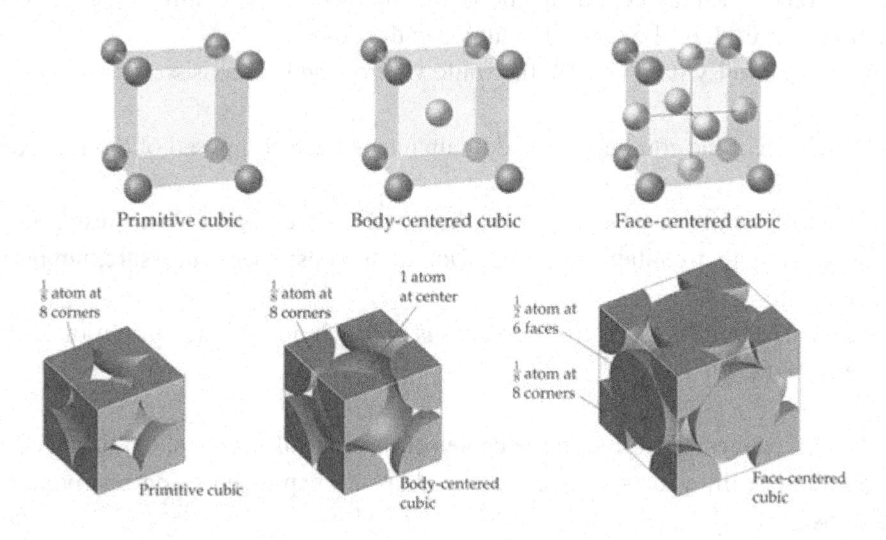

Graciano-Uribe (2021) measured how the porosity of the bed varied with its length, The porosity stabilizes at 0.5 as the bed length increased. Based on the porosity measured experimentally by Graciano-Uribe (2021), it was decided to use the arrangement that had similar porosity. Thus, the chosen arrangement was simple cubic.

Choosing the Type of Contact Between the Spheres

The point of contact between packaging materials is an important region, due to the great difficulty in generating high-quality computational meshes at these points (between particles or between the wall and particle). The geometry of the mesh near the contact points can be distorted, which can lead to convergence and computer simulation problems affecting accuracy. To reduce this effect, four methods were evaluated that aim to change the contact points: gaps (particle shrinkage), overlapping area (particle increase), bridges (cylindrical bridges) and caps (removal of a sphere cover).

According to Ezzatabadipour, et al. (2018) the gap and overlap methods would change the porosity of the bed, negatively affecting the simulation results.

Additionally, comparing bridge and cap methods, Ezzatabadipour, et al. (2018) states that the bridge method better represents the contact between particles and particle-wall for pressure drop. The bridge diameter dimension should be between 16% and 20% of the particle diameter, as smaller diameters result in greater pressure drops in relation to experimental data. Therefore, based on the above, in the present work the bridge contact method was chosen, with a bridge diameter of 16% of the diameter.

Figure 6. Four Methods Analysed for Contact Modification

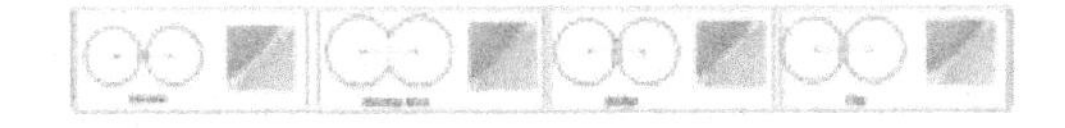

Mesh Generation

The construction of the geometries was based on the work of Kahourzade et al (2013), based on constructions using CS cells with bridge-type contact, adjustments were made to the work of Kahourzade et al (2013) in order to obtain the desired geometries. Table 2 presents the dimensions of the CS cell used as a base.

Figure 7. Cell Dimensions

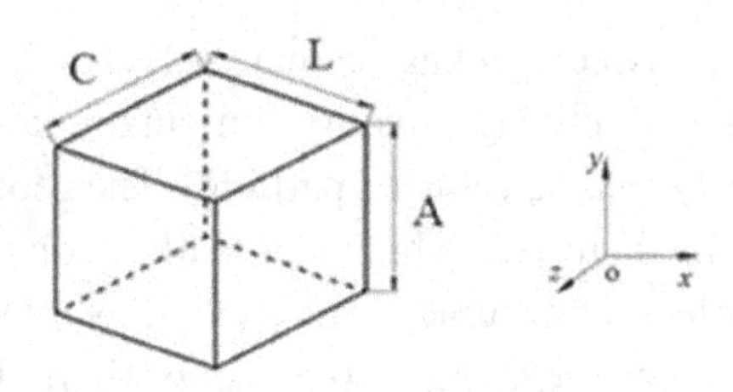

Table 2. CS Unit Cell Dimensions

L (x)	C (z)	A (y)	ε	p
1 cm	1 cm	1 cm	0.48	0.16 cm

To carry out the simulations, it was found that generating a mesh of the complete domain, presented in Figure 2, would require computational cost, so to make the task possible to be performed on a personal computer, only part of the domain was considered. Domain symmetry and periodicity were also adopted. Two cases were analyzed: the first considering a sector of one-eighth of the entire bed, arranged in CS, as shown in the figure below, and the second considering a single CS unit cell.

Figure 8. Front View of the Bed Considered and Top View of the Analyzed Sector

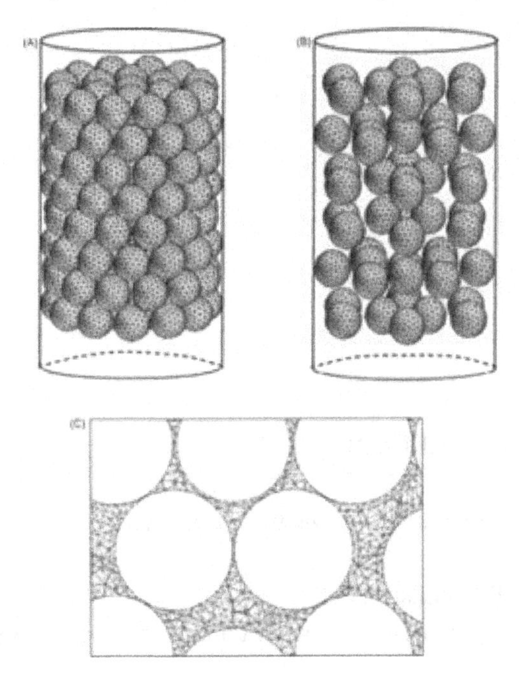

For both cases, the spheres were represented with a dimension close to that of the açaí seed, with a radius equal to 0.5 cm. In order for the periodicity condition to be imposed in Fluent, it was necessary to use the link edges command in Gambit, connecting the nodes associated with the mass input and output areas. The two meshes were generated with tetrahedral elements of approximately 0.025 m in size.

The mesh with a sector of one eighth of the bed had 1298437 volumes and the single cell mesh had 291504 volumes. The meshes can be seen in Figure 10 and Figure 11.

Figure 9. Mesh With Eighth Sector of the Bed and 1298437 Volumes

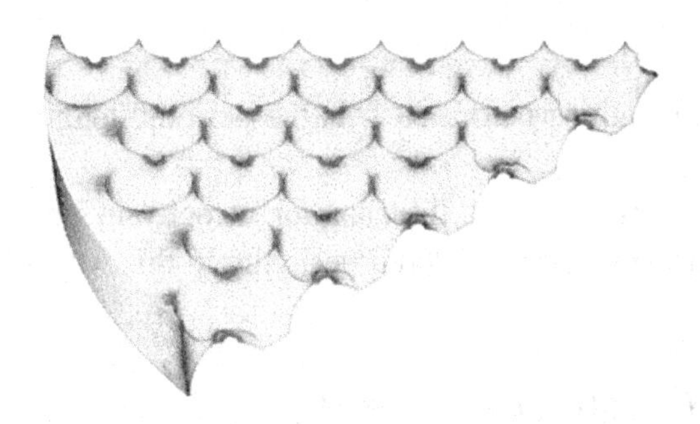

Figure 10. Mesh of CS Single Cell Case, 291504 Volumes

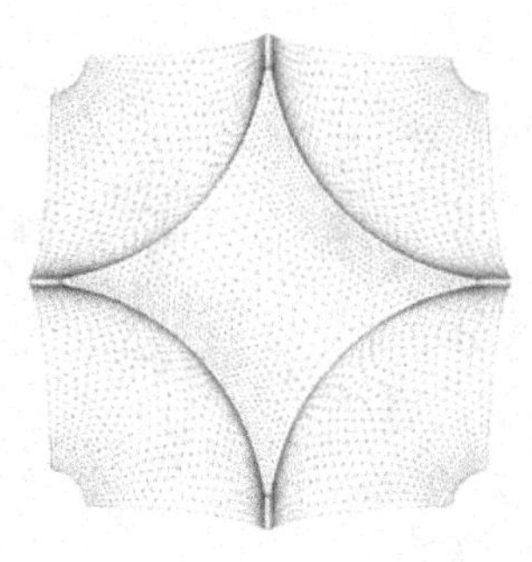

Flow Boundary Conditions

The following boundary conditions were imposed:

- Non-slip on solid wall surfaces:
- Symmetry, which consists of gradient and mass flow in the contour.

The flow was treated as being periodic, that is, the patterns of the analyzed physical geometry and the air flow repeat periodically. When assuming the periodicity of the flow, it is considered that the components of the velocity vector v (v_x, v_y, v_z) are repeated throughout space:

$$v_x(\bar{r}) = v_x(\bar{r} + \bar{L}) = v_x(\bar{r} + \overline{2L}) \quad (8)$$

$$v_y(\bar{r}) = v_y(\bar{r} + \bar{L}) = v_y(\bar{r} + \overline{2L}) \quad (9)$$

$$v_z(\bar{r}) = v_z(\bar{r} + \bar{L}) = v_z(\bar{r} + \overline{2L}) \quad (10)$$

Where $\bar{r}$ is the position vector and $\bar{L}$ is the periodic length vector of the considered domain.

Pressure is not periodic, as is the case for velocity in Eqs. (8), (9) and (10), but the pressure drop can be considered periodic. In this way, Fluent analyzes the periodic flow.

$$p = p\bar{r}) - p(\bar{r} + \bar{L}) = p(\bar{r} + \bar{L}) - p(\bar{r} + \overline{2L}) \quad (11)$$

Thus, the pressure gradient (Pa/m) along the flow was defined as a type of periodic condition.

Average Speed Calculation

The calculation of the average speed was carried out according to Eq. (12).

$$V_m = \frac{1}{\Delta V_f} \int_{V_f} v.dV \tag{12}$$

Where V_f is the volume of fluid contained in a representative elementary volume (ΔV). The integral presented in Eq. (12) is solved numerically by Fluent.

Fluent Simulation Code Adjustments

For the simulations, Fluent was used in a 3ddp (three-dimensional double precision) version, that is, the dimension of the proposed problem (3D) and the type of precision required were specified, in this case double precision. The steady-state laminar flow model was chosen, in this case Fluent solves the equations of conservation of mass and momentum without the transient terms and iteratively.

It was decided to use the segregated equation solution mode, which consumed less memory for the simulations than the coupled equation solution method. Thus, the method to solve the equations that describe the flow was SIMPLE (Semi-Implicit Method for Pressure-Linked Equations). SIMPLE uses a relationship between speed and pressure corrections to reinforce conservation of mass and obtain the pressure field.

A second order upwind scheme was used in the momentum conservation equations. The second order scheme is necessary in the aforementioned equations to minimize the so-called numerical diffusion, which is inherent to the discretization scheme used. The upwind scheme generates numerical diffusion, which is bad, but numerical oscillations are avoided. The PRESTO (PREssure STaggering Option) interpolation scheme was used in the pressure equation.

The convergence criterion adopted was that the normalized residuals should be less than 10-5.

RESULTS AND ANALYSIS

In this part of the work, the results obtained through numerical simulation of the flow through the packed bed of spheres will be presented. A simulation for each pressure gradient experimentally measured by Graciano-Uribe (2021) was performed.

For the two simulated cases, the following conditions were adopted:

- There is no heat transfer during the process.
- The flow is laminar and in steady state.
- Pressure differences per unit length are used as input data.
- Properties are evaluated at a temperature of 298 K.

The average velocity values calculated for each adjusted pressure gradient are presented in Tab. (3) and Tab. (4).

Table 3. Average Speed Values Obtained for a CS Cell

$\Delta P/\Delta L$(kPa/m)	V_m (m/s)
1	3,501
0.5	2,367
0.25	1,540
0.125	0.993
0.0625	0.692
0.022	0.301

Table 4. Average Velocity Values Obtained for One-Eighth of Bed

$\Delta P/\Delta L$(kPa/m)	V_m (m/s)
1	4.58
0.5	3.24
0.125	1.58
0.005	0.28

The results are also presented graphically in Figures 12 and 13. The experimental data for beds with lengths of 1.5 D and 3.5 D were considered, with D being equal to the bed diameter, 0.15 m. For the simulated beds, an infinite length is considered, a product of the flow periodicity condition. Pressure drop values are expressed in units of pressure per unit of length (kPa/m).

Considering that the bed is infinite, the simulated curve should be to the right of all experimental data, which is not the case. The simulated bed is more permeable than the real bed, which should not have happened, as the porosity in this case was 0.48, a value lower than that measured experimentally, which was around 0.5. It is known that the more porous the bed, the less resistance the fluid will encounter to flow, using less energy along its length. It is concluded, therefore, that representing the bed by a single cell that repeats periodically is not an adequate representation of the packed bed in the work of Graciano-Uribe (2021). If the diameter of the experimental bed was much larger than the diameter of the bed spheres, the single-cell approximation could be more successful.

Figure 11. Velocity Versus Pressure Gradient

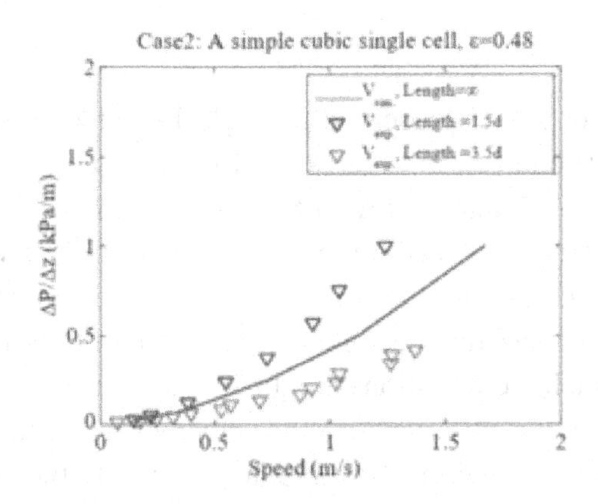

The results considering an eighth of a bed are presented in Figure 19. These were closer to the experimental data. In this case, the simulated bed is less permeable than the experimental bed, this reflects exactly what was expected from a bed longer than the experimental one. The simulation bed represents a packed bed of infinite length. If the experimental bed were larger, perhaps the agreement could be even better between the experimental and the simulated bed. Thus, the domain of one-eighth better represents the flow through a packed bed of spheres.

Figure 12. Velocity Versus Pressure Gradient

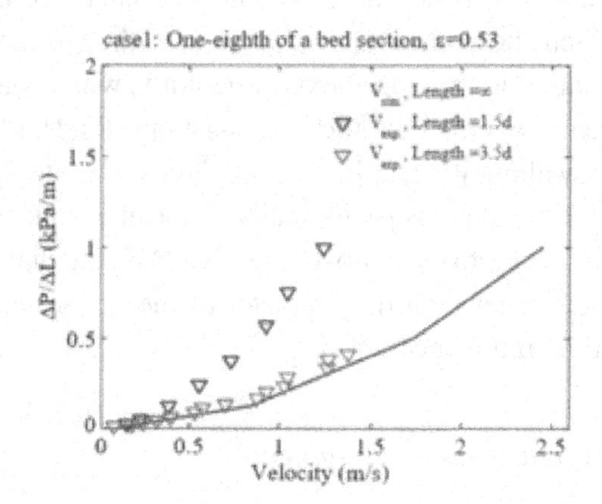

The flow behavior can be observed through the velocity and pressure fields, throughout the entire unit cell in Figures 13, 14, 15, 16 and 17. To generate these figures, an xy plane was created cutting the volume in half. Note the behavior of the flow upstream and downstream of the variation in area in the center of the cell. When the fluid travels through the central region of the cell, the increase in area induces recirculation zones that become more accentuated with the increase in the pressure gradient. Such recirculations are directly associated with the loss of flow energy within the cell. It is observed that the velocities inside the unit cell in Figure 13 and Figure 14 present similar behaviors, however with the increase in pressure in Figure 15 the air recirculation intensifies and causes a considerable increase in velocity in the velocity contours shown.

Figure 13. Velocity Contours m/s for a Pressure Gradient of 22 Pa/m

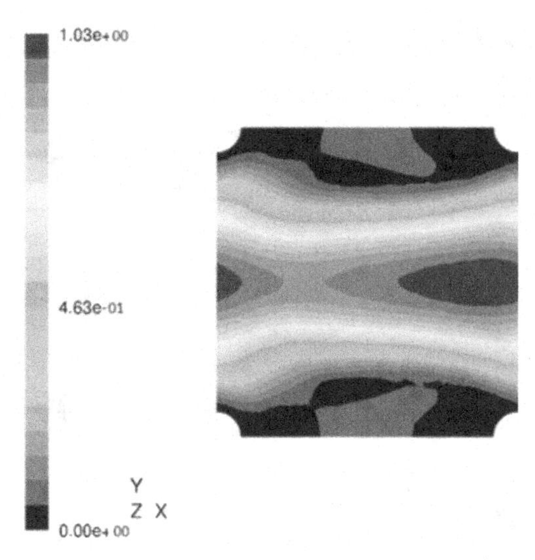

Figure 14. Velocity Contours m/s for a Pressure Gradient of 100 Pa/m

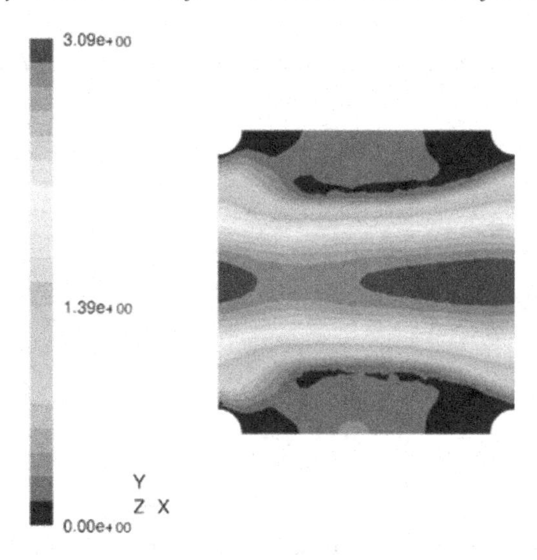

Figure 15. Velocity Contours m/s for a Pressure Gradient of 125 Pa/m

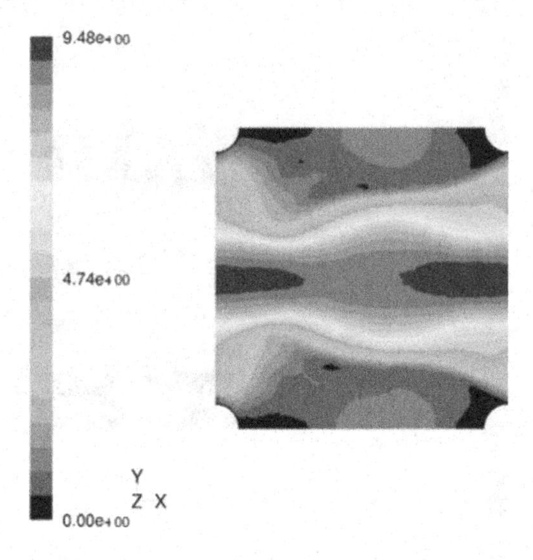

Figure 16 shows an approximation of the recirculation zone. As the flow speed increases, these zones increase.

Figure 16. Zoom into the Recirculation Zone Indicated by the Blue Vectors

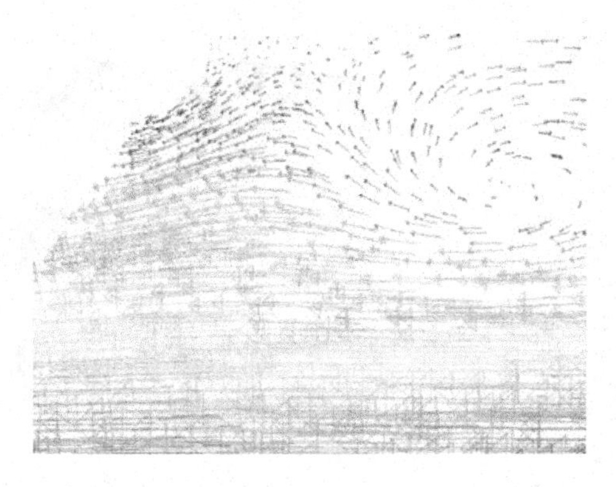

In Figure 17 we can see the intense pressure gradients in the contact regions between the spheres.

Figure 17. Pressure Contour in Pa for the CS Cell

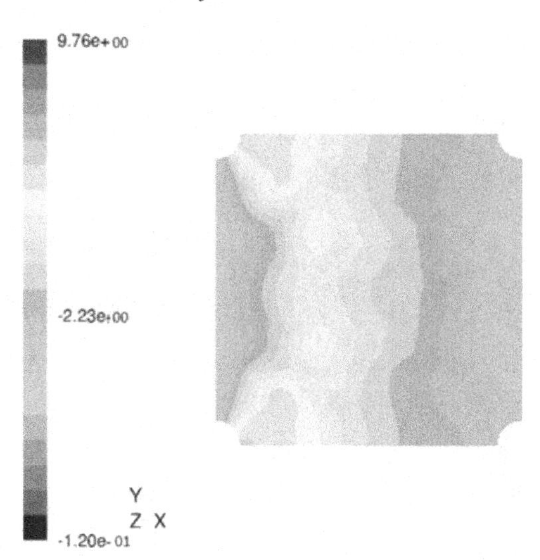

In the case of the one-eighth sector, some differences were observed. An increase in speed was observed in the region close to the wall due to the existence of voids in that region, as can be seen in Figures 18 and 19. It was not possible to fill all the spaces in this configuration. The simulation results indicated that this configuration should be the one that best represents the flow through the packed bed.

Figure 18. Side View of the Velocity Contour for the One-Eighth Sector

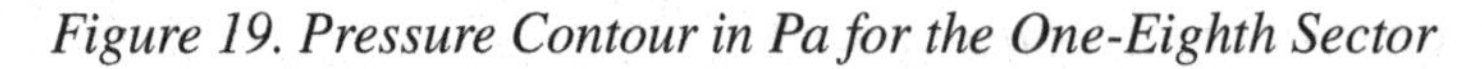

Figure 19. Pressure Contour in Pa for the One-Eighth Sector

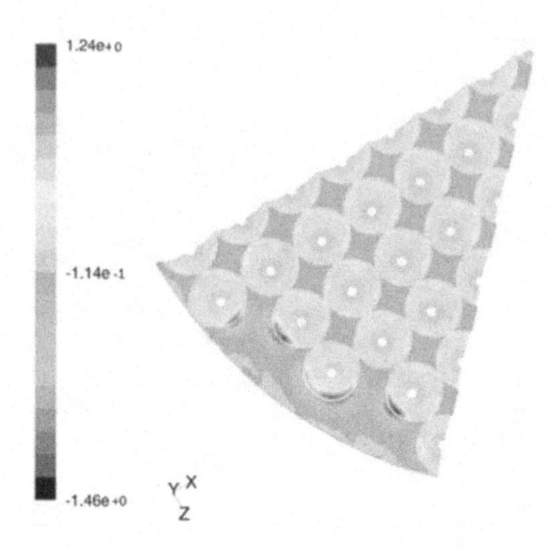

CONCLUSION

In this work, a modelling of the flow through a packed bed of spheres was presented. A review on the subject was presented and the packed bed was appropriately described. In the modelling part, the equations that describe the flow of a fluid, the techniques used to solve the flow and a brief explanation of the geometry and meshes generated by Gambit were presented.

The case study was presented, defining the problem that should be solved. The bed represented with a single CS cell did not approximate the experimentally measured velocities well. It was found that the packed bed was more permeable than the real bed, which should not have happened, as the porosity in this case was 0.48, which is a value lower than that measured experimentally, which was around 0. 5. What can be concluded is that representing the packed bed of Graciano-Uribe 's (2021) work by a single cell that repeats periodically is not a very accurate representation.

The results obtained with an eighth of a bed were closer to the experimental data. The simulated bed was less permeable than the experimental bed, which was exactly what was expected, as the simulated bed represented a bed of infinite length. If the experimental bed were larger, perhaps the agreement would be even better between the experimental and the simulated bed. In such a way, the one-eighth domain represented much better what actually occurred through the Graciano-Uribe (2021) packed bed.

REFERENCES

Barton, N. G. (2013). Simulations of air-blown thermal storage in a rock bed. *Applied Thermal Engineering*, 55(1-2), 43–50. 10.1016/j.applthermaleng.2013.03.002

Cascetta, M., Cau, G., Puddu, P., & Serra, F. (2016). A comparison between CFD simulation and experimental investigation of a packed-bed thermal energy storage system. *Applied Thermal Engineering*, 98, 1263–1272. 10.1016/j.applthermaleng.2016.01.019

Eymard, R., Thierry, G., & Herbin, R. (2000). Finite volume methods. *Finite Volume Methods.*, 7, 713–1018. 10.1016/S1570-8659(00)07005-8

Ezzatabadipour, M., & Zahedi, H. (2018). Simulation of a fluid flow and investigation of a permeability-porosity relationship in porous media with random circular obstacles using the curved boundary lattice Boltzmann method. *The European Physical Journal Plus*, 133(11), 464. 10.1140/epjp/i2018-12325-2

Graciano-Uribe, J., Pujol, T., Puig-Bargués, J., Duran-Ros, M., Arbat, G., & Ramírez de Cartagena, F. (2021). Assessment of Different Pressure Drop-Flow Rate Equations in a Pressurized Porous Media Filter for Irrigation Systems. *Water (Basel)*, 13(16), 2179. 10.3390/w13162179

Kahourzade, S., Mahmoudi, A., Gandomkar, A., Rahim, A., Ping, H. W., & Uddin, M. N. (2013). Design optimization and analysis of AFPM synchronous machine incorporating power density, thermal analysis, and dback-EMF THD. *Electromagnetic Waves*, 136, 327–367. 10.2528/PIER12120204

Keyser, M. J., Conradie, M., Coertzen, M., & Vandyk, J. (2006). Effect of Coal Particle Size Distribution on Packed Bed Pressure Drop and Gas Flow Distribution. *Fuel*, 85(10-11), 1439–1445. 10.1016/j.fuel.2005.12.012

Nemec, D., & Levec, J. (2005). Flow Through Packed Bed Reactors: 1. Single-Phase Flow. *Chemical Engineering Science*, 60(24), 6947–6957. 10.1016/j.ces.2005.05.068

Pesic, R., & Radoičić, K. (2015). Pressure drop in packed beds of spherical particles at ambient and elevated air temperatures. *Chemical Industry & Chemical Engineering Quarterly*, 21(3), 419–427. 10.2298/CICEQ140618044P

Winter, R., Valsamidou, A., Class, H., & Flemisch, B. (2022). A Study on Darcy versus Forchheimer Models for Flow through Heterogeneous Landfills Including Macropores. *Water (Basel)*, 14(4), 546. 10.3390/w14040546

Chapter 9
Solving Nonlinear Problems

Akram H. Shather
Sulaimani Polytechnic University, Iraq

Mohanad Hatem Shadhar
Al-Iraqia University, Iraq

Harish Chandra Bhandari
Kathmandu University, Nepal

ABSTRACT

Over the years, in parallel with technological revolutions, companies have evolved their products to stay competitive, necessitating the recruitment of highly qualified professionals. Engineers play a crucial role in product creation, utilizing CAD/CAE to develop models closely resembling reality, considering various working conditions. This requires understanding real-world phenomena and integrating diverse numerical and computational resources. With Industry 4.0, manufacturing is becoming increasingly digital, demanding engineers adept in innovative virtual tools like augmented reality, 3D printing, and virtual reality. The education sector is adapting to this shift with the introduction of "School 4.0," emphasizing active student participation, with teachers acting as guides rather than sole instructors. This evolution aims to prepare professionals for the dynamic landscape of modern industry, where adaptability and technological fluency are paramount. It highlights the importance of continuous learning and collaboration in driving innovation and sustainable growth.

DOI: 10.4018/979-8-3693-3964-0.ch009

INTRODUCTION

Scientific knowledge is characterized by seeking explanations about physical, chemical, biological and similar events and objects, according to certain acceptance criteria regarding what can be an explanation, a good explanation or a better explanation. In this process, observations, conjectures, experiments, verifications, refutations, concepts, models, theories are at the essence of the construction of scientific knowledge. In other words, this knowledge is constructed, it depends on the questions asked, the definitions, the metaphors, the models used (Rahman, 2021).

An engineer is a professional qualified to design goods and services in the most varied areas using technical and scientific knowledge inherent to their field of training and activity. Its work tools, which until the middle of the 20th century were used manually, gained more agility and precision with the intervention of computers that assist in analysis and decision-making in projects and control.

Currently, engineering companies are faced with several challenges during their projects. The entire process associated with the development of equipment, product or structure must present a good balance of crucial factors, such as quality, cost and time, and the scenario observed today brings several difficulties in achieving this balance. For example, the complexity of products is increasing, resources (both human and raw materials) are scarcer and the deadlines for completing projects and launching products are shorter (Javaid et al., 2022).

Engineering relies on specialized software for developing new products. This technology is known as CAE and encompasses a whole series of systems that help professionals from the analysis of basic physics to more complex systems. Despite simplifying the project development process, their correct operation requires an engineer who has knowledge in physical sciences and the abstraction capacity to create a computational model from a real product (Dodgson et al., 2007). These days, a problem that university instructors who teach solid modelling using CAD (Computer Aided Design) software usually face is the diversity of their students when it comes to their knowledge of resources and tools. Given that they frequently work on CAD projects, the students in the class frequently have a great deal of knowledge and experience in the subject. On the other hand, most students do not have access to these materials or have not used them at all. Furthermore, relying on tutorials and a linear use of resources for classes is no longer useful in the modern day, as a variety of websites, blogs, online forums, and social networks can offer this service without the need for an instructor.

Demonstrate the importance of in-depth study of modeling and simulation using numerical resources in engineering courses, more specifically in the Mechanical Engineering course, through a simulation restricted to the analysis of air flow in an aerodynamic wing, using a CFD (Computation Fluid Dynamics). This study is very common in the development of airplanes in general, automobiles and high-performance boats. Finally, suggest a learning model, based on the principles of Active Methodology, seeking a more efficient dynamic in the process of teaching simulations using Numerical Methods and Mechanical Projects.

According to the quote from Islam (Pranto & Inam, 2020), CFD is a powerful numerical tool that can be used in two aspects: Research and Project. In research, CFD is used primarily as a comparison tool, between numerical results and other types of results, analytical or experimental. In terms of design, CFD is used in practically all branches of engineering, as with the appropriate physical and discretization models, any type of flow can be simulated and viable results obtained.

This chapter interested in emphasizing simulations using Numerical Methods aimed at product development and proving that the expansion in the learning process of this technique is mandatory in Engineering training, taking as an example application in the scope of Mechanical Engineering, where employability of these resources is extremely predominant. It is very important to be aware that this type of work is not just about inserting input data, executing the calculation command and collecting output data; there is an extensive foundation that needs to be explored by the student, uniting the concepts absorbed from the basic cycle and the computational devices contained in the software.

THEORETICAL FOUNDATION

Physical Principles

Next, the main physical concepts that integrate a flow analysis over a submerged solid will be mentioned, in addition to the technique used for experimental analyses, making it possible to explore prototypes on a reduced scale. In this case, 3D printing has been a great ally in obtaining these objects to search for data that can direct the researcher in the creation of theoretical computational models.

Differential Approach

In the study of dynamics in the field of Fluid Mechanics, there are two distinct approaches aimed at particular cases of analysis: Integral approach and differential approach.

According to (Merci, 2016), the first scenario involves examining a spatial region as the fluid moves through it. Equation (1) is a general method for stating that the rate at which a specific characteristic of system N changes is equal to the combined rate of change of the amount of property N within the control volume and the rate at which property N exits the control volume's surface.

$$\frac{dN}{dt}\bigg|_{system} = \frac{\partial \int_{VC} \eta\rho d\forall}{\partial t} + \int_{SC} \eta\rho \, \vec{V} d\vec{A} \tag{1}$$

On the other hand, according to the same author, the differential approach is used when the interest is to analyze the flow in detail. This type of analysis is based on infinitesimal systems and control volumes. Fluid elements are subject to two types of forces: field forces and surface forces. Upon examination of Figure 1, it can be observed that solely the x component of the force dm and volume dV, denoted as dxdydz, are responsible for generating surface forces in the x direction.

Figure 1. Stresses acting on the x-axis in a differential element

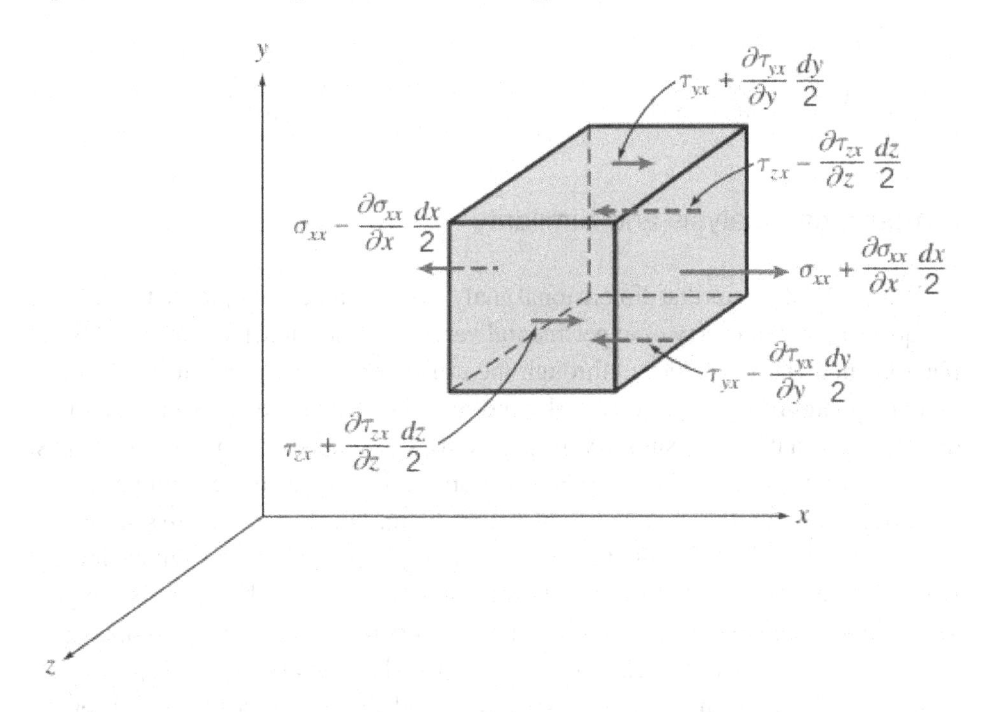

Introducing the equation of acting efforts described in Equation (2), multiplying the tensions by the area differential, in Newton's Second Law, we have the equation of motion with respect to the x axis. Placing tensions as a function of velocity gradients and fluid properties, we reach the most coherent solution to this equation of motion, globally known as the Navier-Stokes Equation.

$$dF|_x = \left(pgx + \frac{\partial \sigma xx}{\partial x} + \frac{\partial ryx}{\partial y} + \frac{\partial \tau zx}{\partial z}\right) dxdydz \tag{2}$$

$$dF|_x = dm\left(\frac{du}{dt}\right) = dm\left(u\frac{\partial u}{\partial x} + v\frac{\partial u}{\partial y} + w\frac{\partial u}{\partial z} + \frac{\partial u}{\partial t}\right) \tag{3}$$

$$\rho\left(u\frac{\partial u}{\partial x} + v\frac{\partial u}{\partial y} + w\frac{\partial u}{\partial z} + \frac{\partial u}{\partial t}\right) = \rho gx - \frac{\partial p}{\partial x} + \mu\left(\frac{\partial^2 u}{\partial x^2} + \frac{\partial^2 u}{\partial y^2} + \frac{\partial^2 u}{\partial z^2}\right) \tag{4}$$

The same idea is applied to the y and z directions, described in Equations (5) and (6)

$$\rho\left(u\frac{\partial u}{\partial x} + v\frac{\partial u}{\partial y} + w\frac{\partial u}{\partial z} + \frac{\partial u}{\partial t}\right) = \rho gy - \frac{\partial p}{\partial y} + \mu\left(\frac{\partial^2 v}{\partial x^2} + \frac{\partial^2 v}{\partial y^2} + \frac{\partial^2 v}{\partial z^2}\right) \tag{5}$$

$$\rho\left(u\frac{\partial u}{\partial x} + v\frac{\partial u}{\partial y} + w\frac{\partial u}{\partial z} + \frac{\partial u}{\partial t}\right) = \rho gz - \frac{\partial p}{\partial z} + \mu\left(\frac{\partial^2 w}{\partial x^2} + \frac{\partial^2 w}{\partial y^2} + \frac{\partial^2 w}{\partial z^2}\right) \tag{6}$$

Dimensional Analysis and Similarity

White (2018) posits that dimensional analysis is a technique employed to decrease the quantity and intricacy of experimental variables that impact a specific physical phenomenon. This is achieved through the utilisation of a compression technique. When a phenomenon is influenced by n dimensional variables, the application of dimensional analysis can simplify the problem to a set of k dimensionless variables. The specific number of dimensionless variables, denoted as nk, can range from 1 to 4, depending on the complexity of the problem. Dimensional analysis, despite its primary objective of reducing variables and grouping them in a dimensionless manner, offers various supplementary advantages. One notable aspect is the provision of scaling laws, which facilitate the conversion of data from a compact and cost-effective model into valuable information for a larger and more costly prototype. A similarity relationship between the model and the prototype can be observed when the scaling law is valid. Similarity is achieved when the Reynolds number is identical between the model and the prototype.

Drag Forces and Support Forces

One of the main output variables of an aerodynamic analysis is force. This, arising from the pressure distribution generated by air contact with the object's surfaces, can be favorable or unfavorable for the project. This is the main reason why this variable must be analyzed, aiming to reduce or increase its intensity and the ideal direction of the vector.

According to (Janna, 2020), drag force is the component of the force on the body that acts parallel to the direction of relative movement. This force can be the result of pure friction between the flow and the surface of the plate, so that the air particles tend to slide on the surface and the friction between them forces the body in the opposite direction to the movement; of pure pressure when the flow is perpendicular to the flow surface, so that the air particles tend to collide frontally with the surface of the body, also generating the opposite force to the movement; and pressure and friction acting simultaneously.

According to (Janna, 2020) lift is defined as the component of the fluid force perpendicular to the fluid movement. They can be directed downwards (down force), with the aim of better stability; and upwards (lift force), with the objective of lifting the object providing flight conditions. Generally, down force is more aimed at high-performance land vehicles and lift force is completely related to aircraft wings and aerial vehicles.

Computational Fluid Mechanics

The following items will briefly conceptualize the entire computational extent of a dynamic analysis within Fluid Mechanics, also known worldwide as CFD. This technique combines computational resources to solve numerical methods applied to specific solutions for systems of Partial Differential Equations that describe flows in a given medium.

Mesh Concept and Discretization

A computational mesh can be defined as a set of points and elements that describe an analysis region or geometry. These elements are polygons connected by points (forming nodes) and their faces, which, when joined, must approach the object of study. In each element there is a mathematical concept, generally formed by systems of Differential Equations that are solved by specific methods so that there is continuity between neighbouring elements. The entirety of this process is referred to as discretization. The process of discretization involves approximating a mathematical expression, such as a function or equations in differential or integral form, that is considered to have infinite continuous values within a specific domain. This approximation is achieved by transforming the expression into analogous expressions that exclusively prescribe values at a finite number of distinct points or volumes within the domain. Figure 2 illustrates an example of a mesh created for aerodynamic analysis of an airplane.

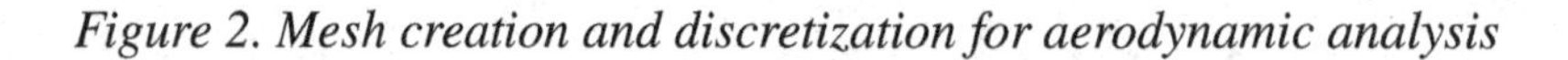

Figure 2. Mesh creation and discretization for aerodynamic analysis

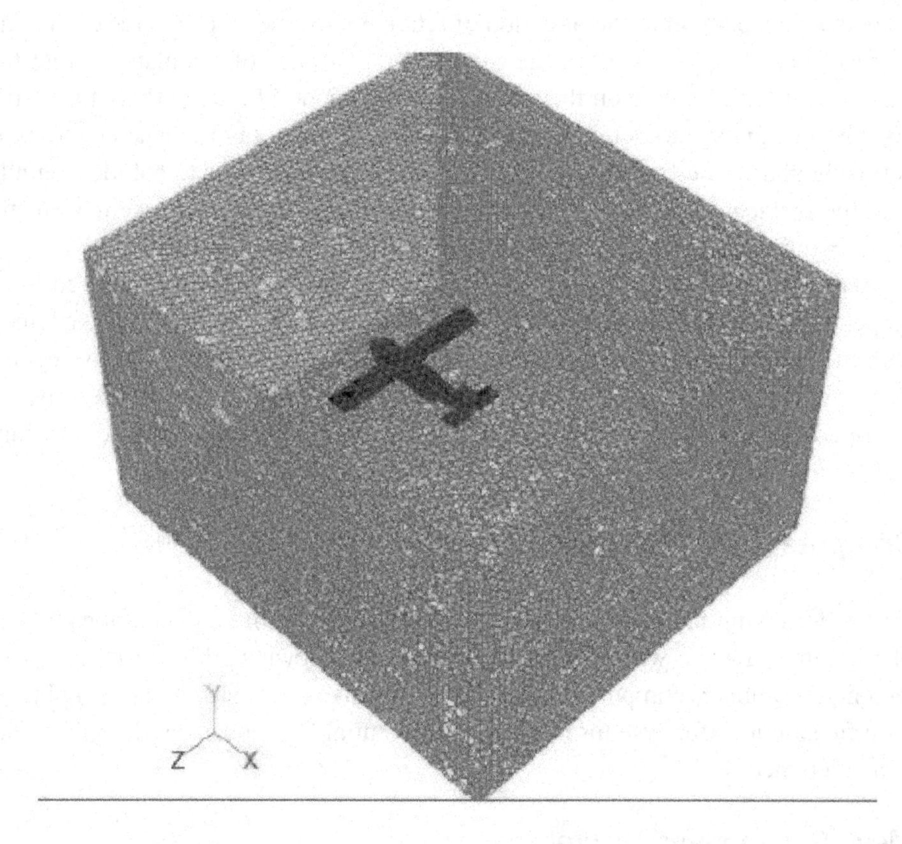

Finite Volume Method

Among the numerical resolution methods for systems of Differential Equations, the method addressed in this study is the Finite Volume Method. The need to solve the Navier-Stokes Equations implies the use of specific resources, while it is also directed to solving cases with this profile.

Analytical solutions of partial differential equations pertain to particular scenarios where expressions provide a continuous variation of the dependent variables across the entire domain. On the other hand, numerical solutions have the capability to provide precise answers at distinct points within the domain (Booker & Nec, 2019).

(Kalies & Do, 2023) asserts that a significant characteristic of this approach is its adherence to the principles of conservation, namely mass, momentum, and energy. These principles serve as the foundation for mathematical modelling in the field of continuum mechanics. Consequently, the equations derived through the volume method inherently uphold these principles. Limited. The methodology is not restricted solely to fluid mechanics problems, but rather encompasses a broader range of steps, which are as follows:

1. The domain is decomposed into control volumes.
2. The integral conservation equations for each control volume are formulated.

The integrals should be approximated numerically.

4. Utilise the information obtained from the nodal variables to estimate the values of the variables on the faces and their derivatives.
5. The algebraic system obtained should be assembled and solved.

Fluid Flow – Ansys

As previously mentioned, Ansys Student was used to simulate a generic case of air flow over an aerodynamic wing. In particular, Ansys has excellent software in terms of manoeuvrability, with very efficient dynamics and high variability of applications in Engineering projects. Fluid Flow (Fluent) was the extension chosen because it has a more complete approach than CFX, which is also aimed at this purpose.

Fluent software has industrial applications such as physical modelling of flow, turbulence, heat transfer, and reactions. These cover a wide range of fields and applications, including but not limited to combustion in furnaces and aeroplane wings, blood flow in semiconductor manufacturing, clean room and wastewater treatment plant design, bubble columns and oil rigs. Fluent covers many topics, including special models, internal combustion, aeroacoustics, turbomachinery, and multiphase systems, among others.

Experimental Analysis in a Wind Tunnel

A wind tunnel (Figure 3) is an installation designed to produce, in a regular and controlled manner, an air current with the aim of experimentally simulating the effects caused on subsurface bodies. Among the effects, we highlight the measurement of efforts characteristic of aerodynamic flows, such as lift and drag forces and the visualization of vortices (Fadrique Ruano, 2017).

Figure 3. Full-Scale test in a wind tunnel

As full-scale testing, with the construction of prototypes, tends to increase the project cost, many organizations use 3D printing to test products on a reduced scale. The theory of Dimensional Analysis and Similarity allows this type of simplification. This is also a trend in so-called companies of the future, in the most varied sectors. Figure 4 shows an application of this type of resource for aerodynamic analysis in the automotive industry.

Figure 4. 3D printing in aerodynamic tests in the automotive industry

METHODOLOGY

For a mathematical model, whether developed by analytical or numerical methods, to be validated, it must be compared with a set of reliable response data, that is, that it is very close to what actually happens in reality. Experimental tests are widely used for this purpose, but they require the researcher to have a good degree of sensitivity and a certain amount of experience to judge the results obtained. Figure 5 depicts the step-by-step process followed to validate the simulation.

Figure 5. Flowchart for simulation validation

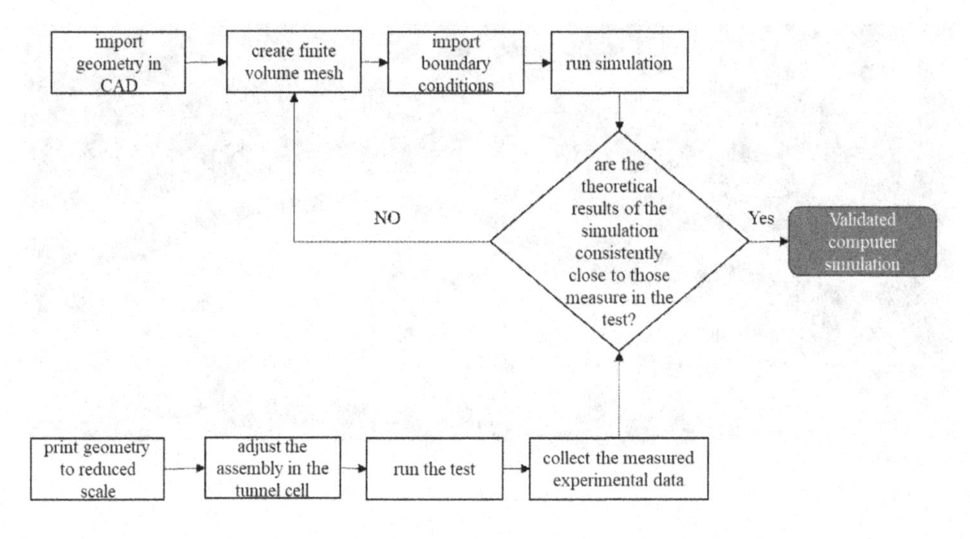

Computational Simulation

In the work described in this article, the field of Fluid Dynamics applied to an aerodynamic analysis of a wing was explored. The student version of Ansys was used, more specifically in the Fluid Flow – Fluent extension.

Geometry Import and Editing

The wing geometry was created in the Catia V5 software and later imported into the IGES format. Design Modeler and Spaceclaim are Ansys extensions available for preparing geometry for simulation. Depending on the CAD tool and the geometric complexity of the component to be simulated, there may be incompatibilities when transferring to Ansys and need to be corrected in either of these two options. As this is an external flow under the surface of a body, it is necessary to limit the test environment with the enclosure tool in the Design Modeler, as shown in Figure 6.

Figure 6. Definition of the simulation environment

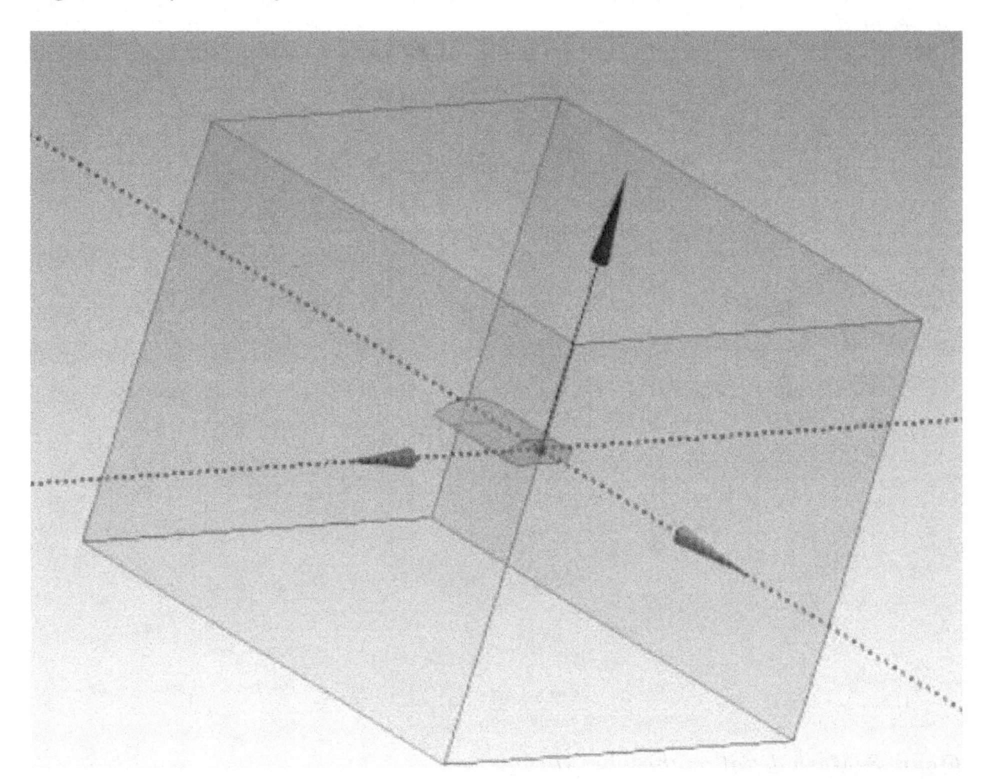

Mesh Generation

The tetrahedral mesh was chosen for the discretization of systems of equations. This type of mesh is a good option for adapting the geometry of the wing in contact with the air that flows through its surfaces. When generating an automatic mesh in the Meshing tab, a good resolution is not always achieved in terms of adaptation to geometry and precision, but there are editing options for these aspects. Among these options, the main ones are: varying the maximum size of the finite volume, its type of geometry and changing the relevance of the mesh. In the case of not very complex geometries, like this one, these resources are not used much, but it is very important to graphically analyze the generated mesh and evaluate possibilities for improvements, this greatly affects the output variables. Observing Figures 7 and 8, in curvature regions, where the insertion of the triangular element is more complex, it is necessary to use smaller elements, adapting them to the geometry and, at the same time, increasing precision in terms of calculation.

Figure 7. Perspective view of the mesh generated in the external environment of the geometry

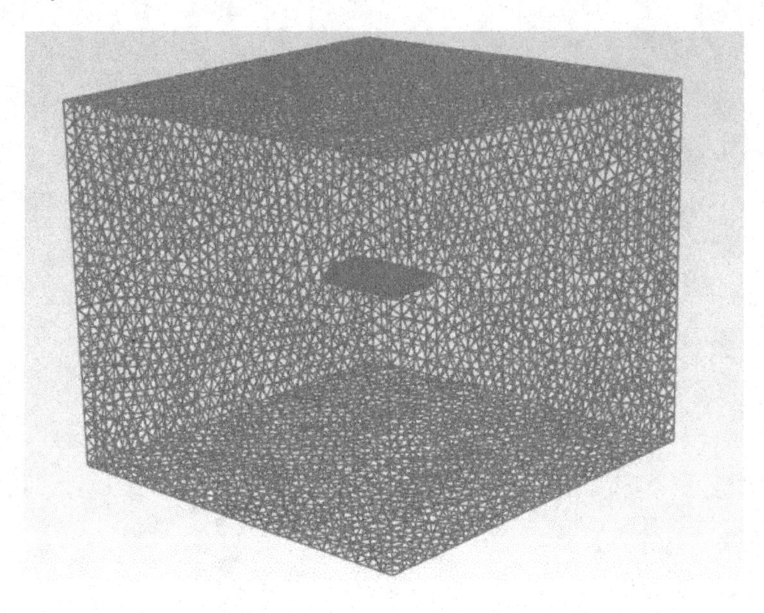

Figure 8. Mesh detail on the wing surface

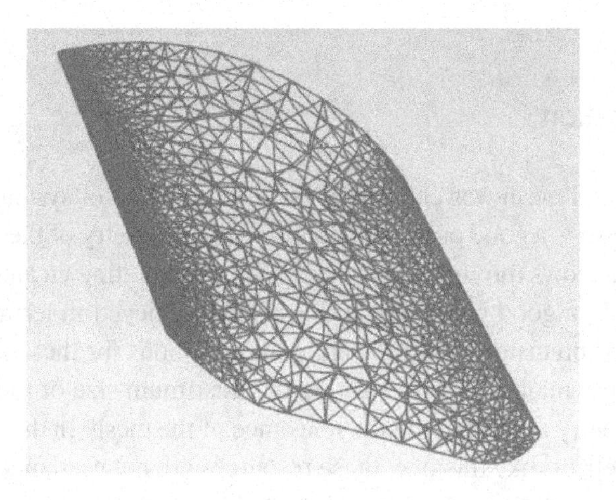

Boundary Conditions

In the process of preparing for the simulation, after creating the mesh, it is necessary to define the boundary conditions of the model. It involves imputing the values of pre-defined input variables to achieve output results. Firstly, it is necessary to define the proportionality between the variables of the theoretical model and the experimental model, based on Dimensional and Similarity Analysis. By definition, when comparing flows in two models with different scales, they have similarity in flow if the Reynolds number is the same for both. With this in mind, it is necessary to seek equality regarding the capacity of the tunnel and the possibilities of manipulation. The real computationally simulated model has three dimensions compared to the model that will be printed in a reduced scale. The flow velocity capacity and absolute pressure capacity of the tunnel are, respectively, 25m/s and 5atm. Therefore, considering the ideal gas law, the density can be varied while the pressure or temperature is varied:

$$\mathrm{Re}\,|_{real} = \mathrm{Re}\,|_{reduced} \quad (7)$$

$$\frac{\rho.v.L}{\mu}\Big|_{real} = \frac{\rho.v.L}{\mu}\Big|_{reduced} \quad (8)$$

$$PV = mRT \therefore \mathrm{R} = \frac{P}{\rho T}\Big|_{real} = \frac{P}{\rho T}\Big|_{reduced} \quad (9)$$

$$T_{real} = T_{reduced} \therefore 5P_{real} = P_{reduced} \tag{10}$$

$$5P_{real} = P_{reduced} \therefore 1/3\, L_{real} = L_{reduced} \therefore 3/5\, v_{real} = v_{reduced} \tag{11}$$

Wind Tunnel Test

In order to validate the computationally achieved solution, the wing geometry will be printed in three dimensions at a scale of 1:3 and subjected to controlled flow in a wind tunnel. To maintain the same Reynolds number, the computationally imputed variables of the real model and the variables of the reduced-scale model must have the intensity as shown in Table 1. The didactic wind tunnel provided by the institution is in accordance with Figure 9.

Figure 9. Didactic wind tunnel

Table 1. Table with the intensity of variables in real and reduced scales

	Real Scale	Reduced Scale
Pressure (Kpa)	101.325	506.625
Speed (m/s)	42	25
Temperature (K)	293	293

RESULTS ANALYSIS AND FUTURE PROPOSALS

Results of Computational Simulation and Comparative With Experimental

Before starting any numerical simulation, it is necessary to define the number of iterations and the stopping criterion, that is, the convergence of the model. Certainly, as we learn in numerical calculation, the lower the error rate, the greater the degree of proximity to an exact solution. However, caution should be taken not to overly define this criterion, which directly compromises the output results. This sensitivity is acquired with experience over time, as the diversity of models that are created and simulated increases. The residual graph defined for the simulation in question is consistent with Figure 10.

Figure 10. Numerical residual graph

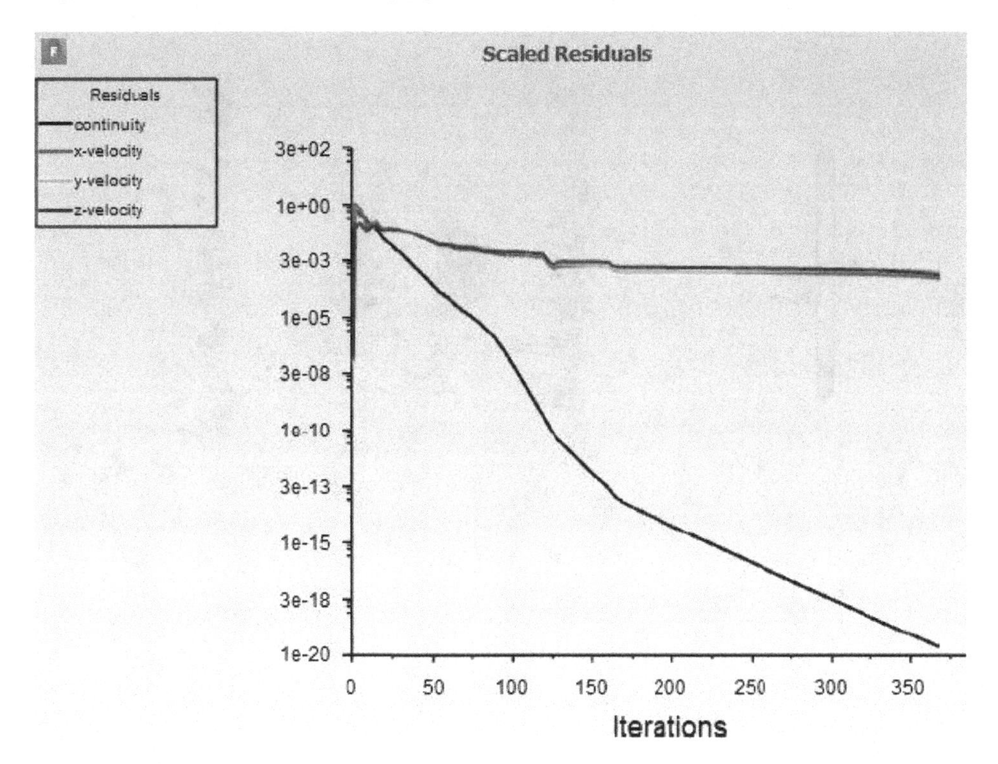

Due to the graphical analysis of speeds in Figure 11, based on the visualization of the directions of the vectors and the intensity indicated in the color legend, it can be stated that the results are not far from what was expected for this case, in some aspects. The direction of the vectors opposite to the movement of the wing and the intensity of the speed has a value close to 40m/s in areas more external to the wing surface layer. However, it was expected that there would be lower values in the contact regions, as the particles come into friction with the surface, generating this reduction. In Figure 12, the parameters for acceptance become more complicated in terms of absolute pressure intensity, which provides drag and lift efforts on the wing. It makes sense that the particles in contact with the surface have a greater intensity of pressure, obviously due to friction with the body, but it is insufficient to consider it a truth. Therefore, an experimental parameter is necessary to compare effort values to check whether there is coherence of results.

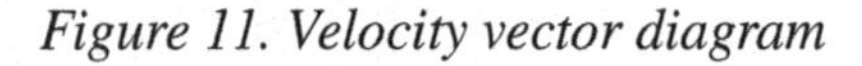

Figure 11. Velocity vector diagram

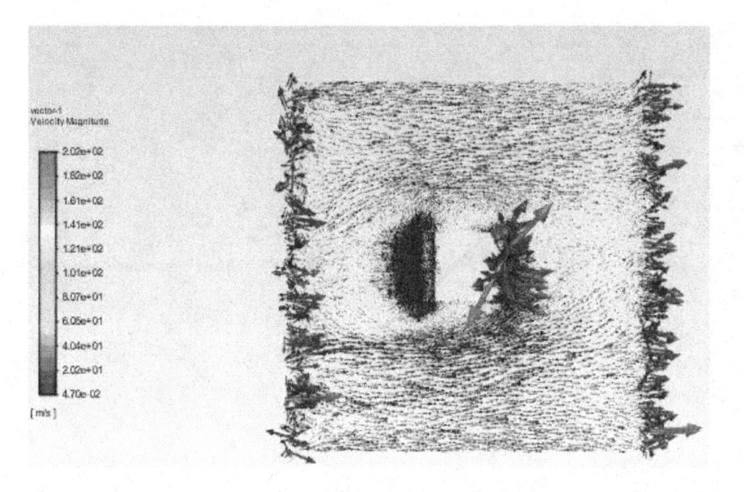

Figure 12. Static pressure vector diagram

Due to filament type incompatibility and other factors related to 3D printing, the wing cannot yet be printed for testing. Observing Figure 13, its structure is breaking before reaching half of the geometry, but there are alternatives to achieve success and, consequently, carry out the test in the tunnel.

Figure 13. 3D printing of the wing

Proposal for Improvement in Teaching

The teaching method proposal is based on Active Methodologies, the main learning model proposal to meet the needs of the industry of the future. Initially, the contribution of collaborators, generally engineers and designers, from partner companies interested in guiding students in their projects is necessary, sharing experiences related to projects they have had in practice throughout their career.

Students of the Mechanical Engineering course in their project disciplines present in the curriculum, Technical Mechanics (4th year) and Construction of Mechanical Machines and Equipment (5th year) will choose a product to be developed since the 4th year, where only the entire planning in terms of related standards and how development would be carried out from the 5th year onwards, creating a control schedule. The Computer Aided Mechanical Design subject, which is currently taught in the 4th year, will provide support in terms of computational tools. It is suggested that the Ansys tool in the student version be covered more in its most common extensions, such as Static Structural, Transient Structural, Fluid Flow (Fluent) and Turbomachienary Fluid Flow, which will certainly be used in the development of the project, in addition to import issues of CAD Software geometry.

From the 5th year onwards, where it is expected that the student will have a certain level of tool mastery and a suggested development schedule in hand, this proposal would be presented to the volunteer collaborator to make initial adjustments and continue with guidance throughout the year, developing the calculations and simulations of this product, defining its entire sizing. However, there would also be a supervisor on the teaching staff to provide opinions on technical issues and guide structural issues in the final work. It is important to highlight that all confidentiality and privacy of the company must be maintained throughout the transition of information between the student and the employee.

Therefore, the student will have to know what their needs are and seek guidance so that their project can be completed successfully. This idea is completely in line with the Active Methodology, where the student is the main agent in the learning process instead of the teacher or advisor. The promise is of benefit to everyone involved:

- Student: In addition to being more self-taught in terms of studies and work, they may have more chances to work in the future at the partner company, in addition to being more prepared for the industry of the future (Nasution, 2020).
- University: It will have a more positive image in society, providing differentiated training for the professionals it is introducing into the market.
- Partner Company: If the mentored student is hired in the future, they will have a more qualified workforce, in addition to being well regarded in the market and in society, participating in this integrative process.

CONCLUSION

Given what was mentioned throughout the article, it can be seen that the objectives listed were partially achieved. It was possible to obtain a prior understanding of the fundamentals of numerical flow analysis, more precisely focused on aerodynamics as well as the high degree of reliability and precision that it provides to projects, drastically reducing the need for creating prototypes and consequently the cost of projects.

Regardless of the area of application, it is essential to deepen the theoretical concepts that involve the entire simulation that you wish to carry out, from the input data to the boundary conditions, to the preparation and discretization of the meshes and to the interpretation of the results achieved.

Above all, it was an opportune opportunity to propose changes to the course, seeking better training for mechanical engineers graduated from the institution and differentiated preparation for the industrial model trend, where the performance of this professional will be indispensable.

REFERENCES

Booker, K., & Nec, Y. (2019). On accuracy of numerical solution to boundary value problems on infinite domains with slow decay. *Mathematical Modelling of Natural Phenomena*, 14(5), 503. 10.1051/mmnp/2019008

Dodgson, M., Gann, D. M., & Salter, A. (2007). The impact of modelling and simulation technology on engineering problem solving. *Technology Analysis and Strategic Management*, 19(4), 471–489. 10.1080/09537320701403425

Fadrique Ruano, G. (2017). *Design of a ventilation system for fan testing*. Academic Press.

Janna, W. S. (2020). *Introduction to fluid mechanics*. CRC press.

Javaid, M., Haleem, A., Singh, R. P., Suman, R., & Gonzalez, E. S. (2022). Understanding the adoption of Industry 4.0 technologies in improving environmental sustainability. *Sustainable Operations and Computers*, 3, 203–217. 10.1016/j.susoc.2022.01.008

Kalies, G., & Do, D. D. (2023). Momentum work and the energetic foundations of physics. IV. The essence of heat, entropy, enthalpy, and Gibbs free energy. *AIP Advances*, 13(9), 095126. 10.1063/5.0166916

Merci, B. (2016). Introduction to fluid mechanics. *SFPE Handbook of Fire Protection Engineering*, 1–24. SFPE.

Nasution, M. K. (2020). *Industry 4.0*. Academic Press.

Pranto, M. R. I., & Inam, M. I. (2020). Numerical Analysis of the Aerodynamic Characteristics of NACA4312 Airfoil. *Journal of Engineering Advancements*, 1(02), 29–36. 10.38032/jea.2020.02.001

Rahman, M. M. (2021). Understanding Science and Preventing It from Becoming Pseudoscience. *Philosophy (London, England)*, 9(3), 127–135.

Chapter 10
Time-Dependent Problems:
Temporal Topological Spaces

Riad Al-Hamido
Alfurat University, Syria

ABSTRACT

In this chapter, the authors study for the first time in the world the notion of new types of temporal open (closed) sets such as temporal semi open (closed) sets, temporal pre open (closed) sets, quasi temporal semi open (closed) sets, and quasi temporal pre open (closed) sets. Also, they studied the relationship between them and temporal open (closed) sets in the temporal topological space. They also created some properties that are achieved in temporal topological space, but are not achieved in topological space or vice versa. Moreover, they prove some remarks, theorems in temporal topological space.

INTRODUCTION

A. Mhemdia et al. (2024) defined a new class of topological space by using temporal topology, which named temporal topological spaces.

As a generalization of topological space, the supra topological space was introduced by A.S. Mashhour et al. (1983). For more information about supra topological space see Al-Shami, (2017a, 2017b, 2018) and Al-Shami et al. (2020). Also, as another generalization of the topological space, Adel. M. AL-Odhari (2015) defined infra topological space. The final space inspired researchers to work on generalizing it, as R.K. Al-Hamido (2023) generalized it to the infra bi-topological space. But the important generalization is the one made by the researchers G. Jayaparthasarathy

DOI: 10.4018/979-8-3693-3964-0.ch010

et al. (2023) where they introduced the concept of neutrosophic supra topological space in 2019, by using the neutrosophic sets.

The concept of neutrosophic bi-topological spaces was introduced by R.K. Al-Hamido (2019) as an extension of neutrosophic topological spaces in 2019. This concept has been studied in Ozturk and Ozkan (2019). Also, The concept of neutrosophic crisp bi-topological spaces was introduced by R.K. Al-Hamido (2018) as an extension of neutrosophic crisp topological spaces in 2018.

In this paper, we defined we study for the first time in the world the notion of new types of temporal open (closed) sets such as temporal semi open (closed) sets, temporal pre open (closed) sets, quasi temporal semi open (closed) sets, and quasi temporal pre open (closed) sets. Also, we studied the relationship between them and temporal open (closed) sets in the temporal topological space. We also studied its basic characteristics.

PRELIMINARIES AND SOME PROPERTIES

In this section, we recall some concepts about temporal topological space from Mhemdia (2024).

Definition 2.1.

Let $\chi \neq \emptyset$ be a set and map $\mathscr{F}: R_+ \wp(\wp(\chi)$ then $(\chi, \mathscr{F})$ called a temporal topological space (TTS) if $\mathscr{F}(\lambda)$ is a topology on χ, for every $\lambda \in R_+$.

In other words:

$(\chi, \mathscr{F})$ called a temporal topological space if for every $\lambda \in R_+$,

1. $\chi, \emptyset \mathscr{F}(\lambda)$;

2. $\mathfrak{P}_1, \mathfrak{P}_2 \mathscr{F}(\lambda)$ $\mathfrak{P}_1 \mathfrak{P}_2 \mathscr{F}((\lambda))$.

3. $(\mathfrak{P}_i)_{iI}$ $\mathscr{F}(x)$ $(\mathfrak{P}_i)_{iI}$ $\mathscr{F}((\lambda))$.

Definition 2.2.

Let $(\chi, \mathscr{F})$ a TTS on χ, then every sub set Q in $\mathscr{F}$ is called temporal open set (TOS) if$Q\mathscr{F}((\lambda))$ for every $(\lambda)\in R_+$.

The set of all TOS is denoted by TOS(χ).

Definition 2.3.

Let $(\chi, \mathscr{F})$ a TTS on χ, then every sub set Q in $\mathscr{F}$ is called temporal closed set (TCS) if$Q\,[\,\mathscr{F}(\lambda)]^{\,c}$ for every $\lambda\in R_+$.

The set of all TCS is denoted by TCS(χ).

Definition 2.4.

Let $(\chi, \mathscr{F})$ a temporal topological space on χ, then every sub set Q in $\mathscr{F}$ is called quasi temporal open set (QTOS) if there exist some $\lambda\in R_+$ such that $Q\mathscr{F}(\lambda)$.

The set of all QTOS is denoted by QTOS(χ).

Remark 2.5.

Every TOS in TTS (χ, T) is QTOS, but the converse is not necessary true.

Definition 2.6.

Let $(\chi, \mathscr{F})$ a TTS on χ, then every sub set Q in $\mathscr{F}$ is called quasi temporal closed set (QTCS) if$Q\,[\,\mathscr{F}(\lambda)]^{\,c}$ for some $\lambda\in R_+$.

The set of all QTOS is denoted by QTOS(χ).

Definition 2.7.

Let $(\chi, \mathscr{F})$ a TTS on χ, and $\mathfrak{H}$ be a sub set in χ, then

$int_T(\mathfrak{H})$ called a temporal interior of $\mathfrak{H}$ and defined as following:

$$int_T(\mathfrak{H}) = \cup\{\mathscr{B};\mathscr{B} \in TOS(\chi)\ and\mathscr{B}\mathfrak{H}\}.$$

$int_Q(\mathfrak{H})$ called a quasi temporal infra interior of $\mathfrak{H}$ and defined as following:

$$int_Q(\mathfrak{H}) = \cup\{\mathscr{B};\mathscr{B} \in QTOS(\chi)\ and\mathscr{B}\mathfrak{H}\}.$$

Definition 2.8.

Let $(\chi, \mathscr{F})$ a TTS on χ, and $\mathfrak{H}$ be a sub set in χ, then:
$cl_T(\mathfrak{H})$ called a temporal closer of $\mathfrak{H}$ and defined as following:

$$cl_T(\mathfrak{H}) = \{\mathscr{B}; \mathscr{B} \in TCS(\chi) \text{ and } \mathfrak{H}\mathscr{B}\}.$$

$cl_Q(\mathfrak{H})$ called a quasi temporal closer of $\mathfrak{H}$ and defined as following:

$$cl_Q(\mathfrak{H}) = \{\mathscr{B}; \mathscr{B} \in QTCS(\chi) \text{ and } \mathfrak{H}\mathscr{B}\}.$$

Remark 2.9.

Let $(\chi, \mathscr{F})$ a TTS on χ, and $\mathfrak{H}$ be a sub set in χ, then:

1. $\mathfrak{H}$ is TOS, iff $\mathfrak{H} = int_T(\mathfrak{H})$.
2. If $\mathfrak{H}$ is QTOS, then $\mathfrak{H} = int_Q(\mathfrak{H})$.

Remark 2.10.

Let $(\chi, \mathscr{F})$ a TTS on χ, and $\mathfrak{H}$ be a sub set in χ, then:

1. $\mathfrak{H}$ is TCS, iff $\mathfrak{H} = cl_T(\mathfrak{H})$.
2. If $\mathfrak{H}$ is QTCS, then $\mathfrak{H} = cl_Q(\mathfrak{H})$.

NEW CLASSES OF SETS IN TEMPORAL TOPOLOGICAL SPACE (TTS)

In this part, we defined new classes of sets in temporal topological space (TTS) such as temporal semi open sets, In additions, we studied its basic properties, Also, we examined the relationships between these sets.

Definition 3.1.

Let $(\chi, \mathscr{F})$ a TTS on χ, and $\mathfrak{H}$ be a sub set in χ, then we called $\mathfrak{H}$ temporal semi open set (TSOS) if there exist a TOS $\mathfrak{o}$ such that $\mathfrak{o} \subseteq \mathfrak{H} \subseteq cl_T(\mathfrak{o})$.

The set of all TSOS is denoted by TSOS(χ).

Theorem 3.2.

Let $(\chi, \mathscr{F})$ a TTS on χ, and $\mathfrak{H}$ be a sub set in χ, then $\mathfrak{H}$ temporal semi open set (TSOS) iff $\mathfrak{H} cl_T(int_T(\mathfrak{H}))$.

Proof

$\Rightarrow$: let $\mathfrak{H}$ temporal semi open set (TSOS), then there exist a TOS $\mathfrak{o}$ such that $\mathfrak{o} \subseteq \mathfrak{H} \subseteq cl_T(\mathfrak{o}) \ldots (i)$.

But $\mathfrak{o}\ int_T(\mathfrak{H})$ and thus $cl_T \subseteq \mathfrak{o} \subseteq cl_T(int_T(\mathfrak{H})) \ldots (ii)$.

Now from (i) and (ii) we have $\mathfrak{H}\ cl_T(\mathfrak{o})\ cl_T(int_T(\mathfrak{H}))$.

$\Leftarrow$: let $\mathfrak{H}\ cl_T(int_T(\mathfrak{H}))$, then foe $\mathfrak{o} = int_T(\mathfrak{H})$ we have $\mathfrak{o} \subseteq \mathfrak{H} \subseteq cl_T(\mathfrak{o}) \ldots (i)$.

But $\mathfrak{o} int_T(\mathfrak{H})$ and thus $cl_T \mathfrak{o} \subseteq cl_T(int_T(\mathfrak{H})) \ldots (ii)$.

Now from (i) and (ii) we have $\mathfrak{H}\ cl_T(\mathfrak{o}) \subseteq cl_T(int_T(\mathfrak{H}))$. Therefore $\mathfrak{H}\ cl_T(int_T(\mathfrak{H}))$.

Theorem 3.3.

Let $(\chi, \mathscr{F})$ a TTS on χ, the union of tow temporal semi open sets is temporal semi open set.

Proof

$\Rightarrow$: let $\mathfrak{H}_1$,$\mathfrak{H}_2$be a temporal semi open set, then $\mathfrak{H}_1\ cl_T(int_T(\mathfrak{H}_1)) \ldots (i)$ and $\mathfrak{H}_2\ cl_T(int_T(\mathfrak{H}_2)) \ldots (ii)$

Now from (i) and (ii) we have

$$\mathfrak{H}_1\ \mathfrak{H}_2\ cl_T(int_T(\mathfrak{H}_1))cl_T(int_T(\mathfrak{H}_2)) = cl_T(int_T(\mathfrak{H}_1)int_T(\mathfrak{H}_2))$$

$$= cl_T(int_T(\mathfrak{H}_1\ \mathfrak{H}_2)).$$

Therefore $\mathfrak{H}_1\ \mathfrak{H}_2$is temporal semi open set.

Theorem 3.4.

Let $(\chi, \mathscr{F})$ a TTS on χ, if $\{\mathfrak{H}_i\}_{iI}$ be a temporal semi open sets, then $\cup_{i\in I}\mathfrak{H}_i$ is a temporal semi open set.

Proof

Since $\{\mathfrak{H}_i\}_{iI}$ be a temporal semi open sets, then there exist a TOS $\mathfrak{o}_i$ such that $\mathfrak{o}_i \subseteq \mathfrak{H} \subseteq cl_T(\mathfrak{o}_i)$ for $i \in I$.

$$\bigcup_{i\in I}\mathfrak{o}_i \subseteq \bigcup_{i\in I}\mathfrak{H}_i \subseteq \bigcup_{i\in I}cl_T(\mathfrak{o}_i) \subseteq cl_T\bigcup_{i\in I}(\mathfrak{o}_i).$$

Let $\mathfrak{o} = \cup_{i\in I}(\mathfrak{o}_i)$, then $\mathfrak{o} \subseteq \cup_{i\in I}\mathfrak{H}_i cl_T(\mathfrak{o})$
Therefore $\cup_{i\in I}\mathfrak{H}_i$ is temporal semi open set.

Theorem 3.5.

Let $(\chi, \mathscr{F})$ a TTS on χ, if $\mathfrak{H}$ be a temporal semi open sets, and let $\mathfrak{H}\mathscr{H}cl_T(\mathfrak{H})$ then $\mathscr{H}$ is a temporal semi open set.

Proof

Since $\mathfrak{H}$ be a temporal semi open sets, then there exist a TOS $\mathfrak{o}$ such that $\mathfrak{o} \subseteq \mathfrak{H} \subseteq cl_T(\mathfrak{o})$. then $\mathfrak{o} \subseteq \mathscr{H}, cl_T(\mathfrak{H}) cl_T(\mathfrak{o})$ thus $\mathscr{H} \subseteq cl_T(\mathfrak{o})$. therefore $\mathfrak{o} \subseteq \mathscr{H} \subseteq cl_T(\mathfrak{o})$.
Therefore $\mathscr{H}$ is temporal semi open set.

Theorem 3.6.

Let $(\chi, \mathscr{F})$ a TTS on χ, every temporal open set is a temporal semi open set.

Proof

Since $\mathfrak{H}$ be a temporal open sets, then $\mathfrak{H} = cl_T(\mathfrak{H})$. we have $int_T(\mathfrak{H})\, cl_T(int_T(\mathfrak{H}))$ thus $\mathfrak{H}\, cl_T(int_T(\mathfrak{H}))$.
Therefore $\mathfrak{H}$ is temporal semi open set.

Remark 3.7.

Converse of the above theorem not true as shown by the following example

Example 3.8.

Let $\chi = \{a, b, c\}$ be a set and map $\mathscr{F}: R_+ \wp(\wp(\chi))$ then $(\chi, \mathscr{F})$ a temporal topological space, where for every $\lambda \in R_+$, $\mathscr{F}(\lambda)$ is give in the following form:

$\mathscr{F}(\lambda) = \{\emptyset, \chi\}$ if $\lambda \in R_+ - \{0, 1, 2\}$.

$\mathscr{F}(0) = \{\emptyset, \chi, \{a\}, \{a,c\}\}, \mathscr{F}(1) = \{\emptyset, \chi, \{a\}, \{a,b\}\},$

$\mathscr{F}(2) = \{\emptyset, \chi, \{b\}, \{a\}, \{a,b\}\}.$

TOS(χ)= $\{\emptyset, \chi, \{a\}\}$.

TCS(χ)= $\{\emptyset, \chi, \{b,c\}\}$.

QTOS(χ)= $\{\emptyset, \chi, \{a,c\}, \{a,b\}, \{b\}, \{a\}\}$.

QTCS(χ)= $\{\emptyset, \chi, \{a,c\}, \{c,b\}, \{b\}, \{c\}\}$.

Let $\mathfrak{H}_1 = \{a,c\}, \mathfrak{H}_2 = \{a,b\}, \mathfrak{H}_3 = \{b,c\}, \mathfrak{H}_4 = \{a\}, \mathfrak{H}_5 = \{b\}, \mathfrak{H}_6 = \{c\}$

Since $\mathfrak{H}_1 cl_T(int_T(\mathfrak{H}_1)) = cl_T(\{a\}) = \chi$, therefore $\mathfrak{H}_1$ is temporal semi open set.

Since $\mathfrak{H}_2 cl_T(int_T(\mathfrak{H}_2)) = cl_T(\{a\}) = \chi$, therefore $\mathfrak{H}_2$ is temporal semi open set.

Since $\mathfrak{H}_3 cl_T(int_T(\mathfrak{H}_3)) = cl_T(\emptyset) = \emptyset$, therefore $\mathfrak{H}_3$ is not temporal semi open set.

Since $\mathfrak{H}_4 cl_T(int_T(\mathfrak{H}_4)) = cl_T(\{a\}) = \chi$, therefore $\mathfrak{H}_4$ is temporal semi open set.

Since $\mathfrak{H}_5 cl_T(int_T(\mathfrak{H}_5)) = cl_T(\emptyset) = \emptyset$, therefore $\mathfrak{H}_5$ is not temporal semi open set.

Since $\mathfrak{H}_6 cl_T(int_T(\mathfrak{H}_6)) = cl_T(\emptyset) = \emptyset$, therefore $\mathfrak{H}_6$ is not temporal semi open set.

TSOS(χ) $= \{\{a\}, \{a,b\}, \{a,c\}, \chi\}$.

Notice that $\{a,b\}, \{a,c\}$ are temporal semi open set, but $\{a,b\}, \{a,c\}$ temporal open set.

TEMPORAL CLOSED SET IN TTS

In this part, we defined new classes of sets in temporal topological space (TTS) such as temporal semi-closed sets, In additions, we studied its basic properties.

Definition 4.1.

Let $(\chi, \mathcal{F})$ a TTS on χ, and $\mathfrak{H}$ be a sub set in χ, then we called $\mathfrak{H}$ temporal semi closed set (TSCS) if there exist a TCS $\mathfrak{o}$ such that $int_T(\mathfrak{o}) \subseteq \mathfrak{H} \subseteq \mathfrak{o}$.

The set of all TSCS is denoted by TSCS(χ).

Theorem 4.2.

Let $(\chi, \mathcal{F})$ a TTS on χ, and $\mathfrak{H}$ be a sub set in χ, then $\mathfrak{H}$ is temporal semi closed set (TSCS) iff $int_T(cl_T(\mathfrak{H}))\mathfrak{H}$.

Proof

$\Rightarrow$: let $\mathfrak{H}$ temporal semi closed set (TSCS), then there exist a TCS $\mathfrak{o}$ such that $int_T(\mathfrak{o}) \subseteq \mathfrak{H} \subseteq \mathfrak{o} \ldots (i)$.

Since $\mathfrak{H} \subseteq \mathfrak{o}$ then $cl_T(\mathfrak{H}) \subseteq \mathfrak{o}$, thus $int_T(cl_T(\mathfrak{H}) \subseteq int_T(\mathfrak{o}) \ldots (ii)$.

Now from (i) and (ii) we have $int_T(cl_T(\mathfrak{H})\mathfrak{H}$.

$\Leftarrow$: let $int_T(cl_T(\mathfrak{H})) \subseteq \mathfrak{H}$, then for $\mathfrak{o} = cl_T(\mathfrak{H})$ we have $int_T(\mathfrak{o}) \subseteq \mathfrak{H} \subseteq \mathfrak{o}$.

Therefore $\mathfrak{H}$ is temporal semi closed set .

Theorem 4.3.

Let $(\chi, \mathcal{F})$ a TTS on χ, and $\mathfrak{H}$ be a sub set in χ, then $\mathfrak{H}$ is temporal semi closed set (TSCS) iff $\chi - \mathfrak{H}$ is temporal semi open set (TSOS).

Proof

$\Rightarrow$: let $\mathfrak{H}$ temporal semi closed set (TSCS), then $int_T(cl_T(\mathfrak{H}))\mathfrak{H}$. Therefore, $\chi - \mathfrak{H}\chi - int_T(cl_T(\mathfrak{H}))$

$= cl_T(\chi - cl_T(\mathfrak{H})) = cl_T(int_T(\chi - \mathfrak{H}))$. Therefore $\chi - \mathfrak{H}$ is temporal semi open set (TSOS).

$\Leftarrow$: let$\chi - \mathfrak{H}$ is temporal semi open set (TSOS). then$\chi - \mathfrak{H} cl_T(int_T(\chi - \mathfrak{H}))$. so, $\mathfrak{H}\chi - cl_T(int_T(\chi - \mathfrak{H})) = int_T(\chi - int_T(\chi - \mathfrak{H})) = int_T(cl_T(\mathfrak{H}))$.

thus, $int_T(cl_T(\mathfrak{H}))\mathfrak{H}$. Therefore $\mathfrak{H}$ is temporal semi closed set.

Theorem 4.4.

Let $(\chi, \mathscr{F})$ a TTS on χ, the intersection of tow temporal semi closed sets is temporal semi closed set.

Proof

⇒: let $\mathfrak{H}_1$,$\mathfrak{H}_2$be a temporal semi closed set, then $int_T(cl_T(\ \mathfrak{H}_1\ ))\,\mathfrak{H}_1$ ...(*i*) and $int_T(cl_T(\ \mathfrak{H}_2\ ))\ \mathfrak{H}_2$...(*ii*)

Now from (*i*) and (*ii*) we have $int_T(cl_T(\ \mathfrak{H}_1\ ))\ int_T(cl_T(\ \mathfrak{H}_2\ ))\,\mathfrak{H}_1\ \mathfrak{H}_2$...(*i*).

But, $int_T(cl_T(\ \ \mathfrak{H}_1\ \ ))\ \ int_T(cl_T(= int_T(cl_T(\ \mathfrak{H}_1 cl_T(\ \mathfrak{H}_2\ )))\,int_T(cl_T(\mathfrak{H}_1\ \mathfrak{H}_1))$

Which mean

$$int_T(cl_T(\mathfrak{H}_1\ \ \mathfrak{H}_1))\ int_T(cl_T(\ \mathfrak{H}_1\ ))\ int_T(cl_T(\ \mathfrak{H}_2\ ))...(ii).$$

Now from (*i*) and (*ii*) we have $int_T(cl_T(\mathfrak{H}_1\ \ \mathfrak{H}_1))\ \ \mathfrak{H}_1\ \mathfrak{H}_2$

Therefore $\mathfrak{H}_1\ \ \ \mathfrak{H}_2$is temporal semi closed set.

Theorem 4.5.

Let $(\chi, \mathscr{F})$ a TTS on χ, if $\{\mathfrak{H}_i\}_{i\in I}$ be a temporal semi closed sets, then $\cap_{i\in I}\mathfrak{H}_i$ is a temporal semi closed set.

Proof

Since $\{\mathfrak{H}_i\}_{i\in I}$ be a temporal semi closed sets, then there exist a TCS $\mathfrak{o}_i$ such that $int_T(\mathfrak{o}_i) \subseteq \mathfrak{H} \subseteq \mathfrak{o}_i$ for $i \in I$.

$$int_T(\cap_{i\in I}(\mathfrak{o}_i)) \subseteq \cap_{i\in I} int_T(\mathfrak{o}_i) \subseteq \cap_{i\in I}\mathfrak{H}_i \subseteq \cap_{i\in I}(\mathfrak{o}_i)$$

Let $\mathfrak{o} = \cap_{i\in I}(\mathfrak{o}_i)$, then

$$int_T(\cap_{i\in I}(\mathfrak{o}_i)) \subseteq \cap_{i\in I}\mathfrak{H}_i \subseteq \cap_{i\in I}(\mathfrak{o}_i)$$

Therefore $\bigcap_{i \in I} \mathfrak{H}_i$ is temporal semi closed set.

Theorem 4.6.

Let $(\chi, \mathscr{F})$ a TTS on χ, if $\mathfrak{H}$ be a temporal semi closed sets, and let $int_T(\mathfrak{H}) \mathscr{H} \mathfrak{H}$ then $\mathscr{H}$ is a temporal semi closed set.

Proof

Since $\mathfrak{H}$ be a temporal semi closed set, then there exist a TCS $\mathfrak{o}$ such that $int_T(\mathfrak{o}) \subseteq \mathfrak{H} \subseteq \mathfrak{o}$. then $\mathscr{H} \subseteq \mathfrak{o}$, but, $int_T(\mathfrak{o}) \subseteq int_T(\mathfrak{H})$ thus $int_T(\mathfrak{o}) \subseteq \mathscr{H}$. therefore $int_T(\mathfrak{o}) \subseteq \mathscr{H} \subseteq \mathfrak{o}$.

Therefore $\mathscr{H}$ is temporal semi closed set.

Theorem 4.7.

Let $(\chi, \mathscr{F})$ a TTS on χ, every temporal closed set is a temporal semi closed set.

Proof

Let $\mathfrak{H}$ be a temporal closed set, then $\mathfrak{H} = cl_T(\mathfrak{H})$. we have $int_T(cl_T(\mathfrak{H}))cl_T(\mathfrak{H})$ this implies $int_T(cl_T(\mathfrak{H}))cl_T(\mathfrak{H})$.

Therefore $\mathfrak{H}$ is temporal semi closed set.

Remark 4.8.

Converse of the above theorem not true as shown by the following example

Example 4.9.

In example 3.8
$\{b\}, \{c\}$ are temporal semi closed sets, but $\{a,b\}, \{a,c\}$ temporal closed sets.

TEMPORAL PRE CLOSED SET IN TTS

Now we will defined Temporal pre closed set in TTS, and study its main properties.

Definition 5.1.

Let $(\chi, \mathscr{F})$ a TTS on χ, and $\mathfrak{H}$ be a sub set in χ, then we called $\mathfrak{H}$ temporal pre open set (TPOS) if $\mathfrak{H}\ int_T(cl_T(\mathfrak{H}))$.
The set of all TPOS is denoted by TPSOS(χ).

Definition 5.2.

Let $(\chi, \mathscr{F})$ a TTS on χ, then every sub set $\mathfrak{H}$ in χ is called temporal pre closed set in χ if its complement is temporal pre open set in χ.
The set of all TPCS is denoted by TPCS(χ).

Example 5.3.

Let $\chi = \{a, b, c\}$ be a set and map $\mathscr{F}: R_+ \wp(\wp(\chi))$ then $(\chi, \mathscr{F})$ a temporal topological space, where for every $\lambda \in R_+$, $\mathscr{F}(\lambda)$ is give in the following form:

$\mathscr{F}(\lambda) = \{\emptyset, \chi\}$

if $\lambda \in R_+ - \{0, 1, 2\}$.

$\mathscr{F}(0) = \{ \emptyset, \chi, \{a,c\} \}, \mathscr{F}(1) = \{ \emptyset, \chi, \{a,b\} \}, \mathscr{F}(2) = \{ \emptyset, \chi, \{b\}, \{a\}, \{a,b\} \}$.

TOS(χ)= $\{\emptyset, \chi\}$.
QTOS(χ)= $\{\emptyset, \chi, \{a,c\}, \{a,b\}, \{b\}, \{a\} \}$.
If $\mathfrak{H}_1 = \{a,c\}$, *then* $\mathfrak{H}_1 int_T(cl_T(\ \mathfrak{H}_1\ )) = int_T(\ \chi\ ) = \chi$, therefore $\mathfrak{H}_1$ is temporal pre open set.
$\mathfrak{H}_1\ cl_T(int_T(\ \mathfrak{H}_1\ )) = cl_T(\emptyset) = \emptyset$, therefore $\mathfrak{H}_1$ is not temporal semi open set.
$\mathfrak{H}_1 int_T\ (cl_T(int_T(\ \mathfrak{H}_1\ ))) = int_T(cl_T(\ \emptyset\ )) = \emptyset$, therefore $\mathfrak{H}_1$ is not temporal semi open set.

NEW CLASSES OF QUASI SETS IN TEMPORAL TOPOLOGICAL SPACE

In this part, we defined new classes of sets in temporal topological space such as quasi temporal semi open (semi closed) sets, quasi temporal per open (pre closed) sets. In additions, we studied its basic properties, Also, we examined the relationships between these sets.

Definition 6.1.

Let $(\chi, \mathcal{F})$ a TTS on χ, and $\mathfrak{H}$ be a sub set in χ, then we called $\mathfrak{H}$ quasi temporal semi open set (QTSOS) if $\mathfrak{H}\ cl_Q(int_Q(\mathfrak{H}))$.

The set of all QTSOS is denoted by QTSOS(χ).

Definition 6.2.

Let $(\chi, \mathcal{F})$ a TTS on χ, and $\mathfrak{H}$ be a sub set in χ, then we called $\mathfrak{H}$ quasi temporal pre open set (QTPOS) if $\mathfrak{H}\ int_Q(cl_Q(\mathfrak{H}))$.

The set of all QTPOS is denoted by QTPSOS(χ).

Definition 6.3.

Let $(\chi, \mathcal{F})$ a TTS on χ, then every sub set $\mathfrak{H}$ in χ is called quasi temporal (semi, pre) closed set in χ if its complement is quasi temporal (semi, pre) open set in χ.

The set of all QTSCS is denoted by QTSCS(χ).

The set of all QTPCS is denoted by QTPCS(χ).

Example 6.4.

In example 3.8.

If $\mathfrak{H}_1 = \{a,c\}$, *then* $\mathfrak{H}_1 int_Q(cl_Q(\mathfrak{H}_1)) = int_Q(\{a,c\}) = \{a,c\}$, therefore $\mathfrak{H}_1$ is quasi temporal pre open set.

If $\mathfrak{H}_2 = \{a,b\}$, *then* $\mathfrak{H}_2 int_Q(cl_Q(\mathfrak{H}_2)) = int_Q(\chi) = \chi$, therefore $\mathfrak{H}_2$ is quasi temporal pre open set.

$\mathfrak{H}_1\ cl_T(int_T(\mathfrak{H}_1)) = cl_T(\emptyset) = \emptyset$, therefore $\mathfrak{H}_1$ is not temporal semi open set. But, $\mathfrak{H}_1\ cl_T(int_T(\mathfrak{H}_1)) = cl_T(\mathfrak{H}_1) = \mathfrak{H}_1$. Therefore, $\mathfrak{H}_1$is quasi temporal semi open set.

$\mathfrak{H}_1 int_T\ (cl_T(int_T(\mathfrak{H}_1))) = int_T(cl_T(\emptyset)) = \emptyset$, therefore $\mathfrak{H}_1$ is not temporal semi open set.

Remark 6.5.

There are quasi temporal semi open set but not temporal semi open set Such as$\mathfrak{H}_1$ in example 4.6.

Example 6.6.

Let $\chi = \{a, b, c\}$ be a set and map $\mathcal{F}: R_+ \wp(\wp(\chi))$ then $(\chi, \mathcal{F})$ a temporal topological space, where for every $\lambda \in R_+$, $\mathcal{F}(\lambda)$ is give in the following form:

$\mathcal{F}(\lambda) = \{\emptyset, \chi\}$ if $\lambda \in R_+ - \{0, 1, 2\}$.

$\mathcal{F}(0) = \{ \emptyset, \chi, \{a\}, \{a,c\} \}, \mathcal{F}(1) = \{ \emptyset, \chi, \{a\}, \{a,b\} \}$,

$\mathcal{F}(2) = \{ \emptyset, \chi, \{b\}, \{a\}, \{a,b\} \}$.

TOS(χ)= $\{\emptyset, \chi, \{a\}\}$.
TCS(χ)= $\{\emptyset, \chi, \{b,c\}\}$.
QTOS(χ)= $\{\emptyset, \chi, \{a,c\}, \{a,b\}, \{b\}, \{a\} \}$.
QTCS(χ)= $\{\emptyset, \chi, \{a,c\}, \{c,b\}, \{b\}, \{c\} \}$.
Let $\mathfrak{H}_1 = \{a,c\}, \mathfrak{H}_2 = \{a,b\}, \mathfrak{H}_3 = \{b,c\}, \mathfrak{H}_4 = \{a\}, \mathfrak{H}_5 = \{b\}, \mathfrak{H}_6 = \{c\}$
Since $\mathfrak{H}_1 \, int_T(cl_T(\mathfrak{H}_1)) = int_T(\chi) = \chi$, therefore $\mathfrak{H}_1$ is temporal pre open set.
Since $\mathfrak{H}_2 \, int_T(cl_T(\mathfrak{H}_2)) = int_T(\chi) = \chi$, therefore $\mathfrak{H}_2$ is temporal pre open set.
Since $\mathfrak{H}_3 \, int_T(cl_T(\mathfrak{H}_3)) = int_T(\{b,c\}) = \emptyset$, therefore $\mathfrak{H}_3$ is not temporal pre open set.
Since $\mathfrak{H}_4 \, int_T(cl_T(\mathfrak{H}_4)) = int_T(\chi) = \chi$, therefore $\mathfrak{H}_4$ is temporal pre open set.
Since $\mathfrak{H}_5 \, int_T(cl_T(\mathfrak{H}_5)) = int_T(\{b,c\}) = \emptyset$, therefore $\mathfrak{H}_5$ is not temporal pre open set.
Since $\mathfrak{H}_6 \, int_T(cl_T(\mathfrak{H}_6)) = int_T(\{b,c\}) = \emptyset$, therefore $\mathfrak{H}_6$ is not temporal pre open set.
Therefore, TPOS(χ) = $\{ \{a\}, \{a,b\}, \{a,c\}, \chi \}$.
$\mathfrak{H}_1 \, int_Q(cl_Q(\mathfrak{H}_1)) = int_Q(\mathfrak{H}_1) = \mathfrak{H}_1$, therefore $\mathfrak{H}_1$ is quasi temporal pre open set.
$\mathfrak{H}_2 \, int_Q(cl_Q(\mathfrak{H}_2)) = int_Q(\chi) = \chi$, therefore $\mathfrak{H}_2$ is quasi temporal pre open set.
$\mathfrak{H}_3 \, int_Q(cl_Q(\mathfrak{H}_3)) = int_Q(\{b,c\}) = \{b\}$, therefore $\mathfrak{H}_3$ is not quasi temporal pre open set.
$\mathfrak{H}_4 \, int_Q(cl_Q(\mathfrak{H}_4)) = int_Q(\{a,c) = \{a,c\}$, therefore $\mathfrak{H}_4$ is quasi temporal pre open set.

$\mathfrak{H}_5 int_T(cl_T(\ \mathfrak{H}_5\ )) = int_T(\ \{b\}\ ) = \{b\}$, therefore $\mathfrak{H}_5$ is not quasi temporal pre open set.

$\mathfrak{H}_6 int_Q(cl_Q(\ \mathfrak{H}_6\ )) = int_Q(\ \{c\}\ ) = \emptyset$, therefore $\mathfrak{H}_6$ is not quasi temporal pre open set.

Therefore, QTPOS(χ) $= \{\ \{a\}, \{b\}, \{a,b\}, \{a,c\}, \chi\ \}$.

Since $\mathfrak{H}_1 cl_T(int_T(\ \mathfrak{H}_1\ )) = cl_T(\ \{a\}\ ) = \chi$, therefore $\mathfrak{H}_1$ is temporal semi open set.

Since $\mathfrak{H}_2 cl_T(int_T(\ \mathfrak{H}_2\ )) = cl_T(\ \{a\}\ ) = \chi$, therefore $\mathfrak{H}_2$ is temporal semi open set.

Since $\mathfrak{H}_3 cl_T(int_T(\ \mathfrak{H}_3\ )) = cl_T(\ \emptyset\ ) = \emptyset$, therefore $\mathfrak{H}_3$ is not temporal semi open set.

Since $\mathfrak{H}_4 cl_T(int_T(\ \mathfrak{H}_4\ )) = cl_T(\ \{a\}\ ) = \chi$, therefore $\mathfrak{H}_4$ is temporal semi open set.

Since $\mathfrak{H}_5 cl_T(int_T(\ \mathfrak{H}_5\ )) = cl_T(\ \emptyset\ ) = \emptyset$, therefore $\mathfrak{H}_5$ is not temporal semi open set.

Since $\mathfrak{H}_6 cl_T(int_T(\ \mathfrak{H}_6\ )) = cl_T(\ \emptyset\ ) = \emptyset$, therefore $\mathfrak{H}_6$ is not temporal semi open set.

TSOS(χ) $= \{\ \{a\}, \{a,b\}, \{a,c\}, \chi\ \}$.

$\mathfrak{H}_1 cl_Q(int_Q(\ \mathfrak{H}_1\ )) = cl_Q(\ \{a,c\}\ ) = \{a,c$, therefore $\mathfrak{H}_1$ is quasi temporal semi open set.

$\mathfrak{H}_2 cl_T(int_T(\ \mathfrak{H}_2\ )) = cl_Q(\ \{a,b\}\ ) = \chi$, therefore $\mathfrak{H}_2$ is quasi temporal semi open set.

$\mathfrak{H}_3 cl_Q(int_Q(\ \mathfrak{H}_3\ )) = cl_Q(\ \{b\}\ ) = \{b\}$, therefore $\mathfrak{H}_3$ is not quasi temporal semi open set.

$\mathfrak{H}_4 cl_Q(int_Q(\ \mathfrak{H}_4\ )) = cl_Q(\ \{a\}\ ) = \{a,c\}$, therefore $\mathfrak{H}_4$ is quasi temporal semi open set.

$\mathfrak{H}_5 cl_Q(int_Q(\ \mathfrak{H}_5\ )) = cl_Q(\ \{b\}\ ) = \{b\}$, therefore $\mathfrak{H}_5$ is not quasi temporal semi open set.

$\mathfrak{H}_6 cl_Q(int_Q(\ \mathfrak{H}_6\ )) = cl_Q(\ \emptyset\ ) = \emptyset$, therefore $\mathfrak{H}_6$ is not quasi temporal semi open set.

QTSOS(χ) $= \{\ \{a\}, \{b\}, \{a,b\}, \{a,c\}, \chi\ \}$.

TSOS(χ) $= \{\ \{a\}, \{a,b\}, \{a,c\}, \chi\ \}$

Remark 6.7.

There are quasi temporal pre open set but not temporal pre open set Such as $\{b\}$ in example 6.6.

Remark 6.8.

There are temporal semi open set but not quasi temporal semi open set Such as $\{b\}$ in example 6.6.

Example 6.9.

Let $\chi = \{a, b, c\}$ be a set and map $\mathcal{F}:R_+ \wp(\wp(\chi))$ then $(\chi, \mathcal{F})$ a temporal topological space, where for every $\lambda \in R_+, \mathcal{F}(\lambda)$ is give in the following form:

$\mathcal{F}(\lambda) = \{\emptyset, \chi, \{a,b\}, \{b\}\}$ if $\lambda \in R_+ - \{0, 1\}$.

$\mathcal{F}(0) = \{ \emptyset, \chi, \{a\}, \{b\}, \{a,b\} \}$,

$\mathcal{F}(1) = \{ \emptyset, \chi, \{a\}, \{b\}, \{c\}, \{a,b\}, \{a,c\}, \{b,c\} \}$.

TOS(χ)= $\{\emptyset, \chi, \{a,b\}, \{b\}\}$.
TCS(χ)= $\{\emptyset, \chi, \{a,c\}, \{c\} \}$.
QTOS(χ)= $\{\emptyset, \chi, \{a,c\}, \{a,b\}, \{b\}, \{a\}, \{c\}, \{b,c\} \}$.
QTCS(χ)= $\{\emptyset, \chi, \{a,c\}, \{c,b\}, \{a,b\}, \{a\}, \{b\}, \{c\} \}$.
Let $\mathfrak{H}_1 = \{a,c\}, \mathfrak{H}_2 = \{a,b\}, \mathfrak{H}_3 = \{b,c\}, \mathfrak{H}_4 = \{a\}, \mathfrak{H}_5 = \{b\}, \mathfrak{H}_6 = \{c\}$
Since $\mathfrak{H}_1 int_T(cl_T(\mathfrak{H}_1)) = int_T(\mathfrak{H}_1) = \emptyset$, therefore $\mathfrak{H}_1$ is not temporal pre open set.
Since $\mathfrak{H}_2 int_T(cl_T(\mathfrak{H}_2)) = int_T(\chi) = \chi$, therefore $\mathfrak{H}_2$ is temporal pre open set.
Since $\mathfrak{H}_3 int_T(cl_T(\mathfrak{H}_3)) = int_T(\chi) = \chi$, therefore $\mathfrak{H}_3$ is temporal pre open set.
Since $\mathfrak{H}_4 int_T(cl_T(\mathfrak{H}_4)) = int_T(\{a,c\}) = \emptyset$, therefore $\mathfrak{H}_4$ is not temporal pre open set.
Since $\mathfrak{H}_5 int_T(cl_T(\mathfrak{H}_5)) = int_T(\chi) = \chi$, therefore $\mathfrak{H}_5$ is temporal pre open set.
Since $\mathfrak{H}_6 int_T(cl_T(\mathfrak{H}_6)) = int_T(\{c\}) = \emptyset$, therefore $\mathfrak{H}_6$ is not temporal pre open set.
Therefore, TPOS(χ) $= \{ \{a,b\}, \{b\}, \chi \}$.
$\mathfrak{H}_1 int_Q(cl_Q(\mathfrak{H}_1)) = int_Q(\mathfrak{H}_1) = \mathfrak{H}_1$, therefore $\mathfrak{H}_1$ is quasi temporal pre open set.

$\mathfrak{H}_2 int_Q(cl_Q(\mathfrak{H}_2)) = int_Q(\mathfrak{H}_2) = \mathfrak{H}_2$, therefore $\mathfrak{H}_2$ is quasi temporal pre open set.

$\mathfrak{H}_3 int_Q(cl_Q(\mathfrak{H}_3)) = int_Q(\mathfrak{H}_3) = \mathfrak{H}_3$, therefore $\mathfrak{H}_3$ is not quasi temporal pre open set.

$\mathfrak{H}_4 int_Q(cl_Q(\mathfrak{H}_4)) = int_Q(\{a,c) = \{a,c\}$, therefore $\mathfrak{H}_4$ is quasi temporal pre open set.

$\mathfrak{H}_5 int_T(cl_T(\mathfrak{H}_5)) = int_T(\mathfrak{H}_5) = \mathfrak{H}_5$, therefore $\mathfrak{H}_5$ is not quasi temporal pre open set.

$\mathfrak{H}_6 int_Q(cl_Q(\mathfrak{H}_6)) = int_Q(\mathfrak{H}_6) = \mathfrak{H}_6$, therefore $\mathfrak{H}_6$ is not quasi temporal pre open set.

Therefore, QTPOS(χ) = $\{ \{a\},\{b\}, \{c\},\{a,b\},\{a,c\},\{b,c\},\chi \}$.

Since $\mathfrak{H}_1 cl_T(int_T(\mathfrak{H}_1)) = cl_T(\emptyset) = \emptyset$, therefore $\mathfrak{H}_1$ is not temporal semi open set.

Since $\mathfrak{H}_2 cl_T(int_T(\mathfrak{H}_2)) = cl_T(\{a,b\}) = \chi$, therefore $\mathfrak{H}_2$ is temporal semi open set.

Since $\mathfrak{H}_3 cl_T(int_T(\mathfrak{H}_3)) = cl_T(\{b\}) = \chi$, therefore $\mathfrak{H}_3$ is not temporal semi open set.

Since $\mathfrak{H}_4 cl_T(int_T(\mathfrak{H}_4)) = cl_T(\emptyset) = \emptyset$, therefore $\mathfrak{H}_4$ is temporal semi open set.

Since $\mathfrak{H}_5 cl_T(int_T(\mathfrak{H}_5)) = cl_T(\{b\}) = \chi$, therefore $\mathfrak{H}_5$ is temporal semi open set.

Since $\mathfrak{H}_6 cl_T(int_T(\mathfrak{H}_6)) = cl_T(\emptyset) = \emptyset$, therefore $\mathfrak{H}_6$ is not temporal semi open set.

TSOS(χ) = $\{ \{a\},\{a,b\},\{a,c\},\chi \}$.

- QTSOS(χ)= $\{ \chi, \{b,c\} ,\{a,b\}, \{b\} \}$.

$\mathfrak{H}_1 cl_Q(int_Q(\mathfrak{H}_1)) = cl_Q(\{a,c\}) = \{a,c\}$, therefore $\mathfrak{H}_1$ is quasi temporal semi open set.

$\mathfrak{H}_2 cl_T(int_T(\mathfrak{H}_2)) = cl_Q(\{a,b\}) = \{a,b\}$, therefore $\mathfrak{H}_2$ is quasi temporal semi open set.

$\mathfrak{H}_3 cl_Q(int_Q(\mathfrak{H}_3)) = cl_Q(\{c,b\}) = \{c,b\}$, therefore $\mathfrak{H}_3$ is quasi temporal semi open set.

$\mathfrak{H}_4 cl_Q(int_Q(\mathfrak{H}_4)) = cl_Q(\{a\}) = \{a\}$, therefore $\mathfrak{H}_4$ is quasi temporal semi open set.

$\mathfrak{H}_5 cl_Q(int_Q(\mathfrak{H}_5)) = cl_Q(\{b\}) = \{b\}$, therefore $\mathfrak{H}_5$ is quasi temporal semi open set.

$\mathfrak{H}_6 cl_Q(int_Q(\mathfrak{H}_6)) = cl_Q(\{c\}) = \{c\}$, therefore $\mathfrak{H}_6$ is quasi temporal semi open set.

QTSOS(χ) = $\{ \chi, \{a,c\},\{a,b\}, \{b\}, \{a\},\{c\},\{b,c\}\}$.

Theorem 6.10.

Let $(\chi, \mathscr{F})$ a TTS on χ, then:

1. Every TOS is a TPOS.
2. Every TOS is a TSOS.

Proof:

1. Let $\mathfrak{H}$ be a TOS, since $\mathfrak{H}$ be a TOS, then $\mathfrak{H} = int_T(\mathfrak{H}) cl_T(int_T(\mathfrak{H}))$.

So $\mathfrak{H} = int_T(\mathfrak{H})\ int_T cl_T(int_T(\mathfrak{H}))$. Therefore, $\mathfrak{H}\ int_T cl_T(int_T(\mathfrak{H}))\ldots(I)$.
Now, we have $int_T(\mathfrak{H})\ \mathfrak{H}$. So $int_T(cl_T(int_T(\mathfrak{H})))\ int_T(cl_T(\mathfrak{H}))\ldots(II)$. From (I) and (II) we have $\mathfrak{H}\ int_T(cl_T(\mathfrak{H}))$. Therefore $\mathfrak{H}$ is TPOS.

2. In same way to (1).

CONCLUSION

The main goal of this work is to generalize the open (closed) sets in topological space to the temporal open (closed) sets in temporal topological space. In this work, we defined new types of temporal open (closed) sets such as temporal semi open (closed) sets, temporal pre open (closed) sets, quasi temporal semi open (closed) sets, and quasi temporal pre open (closed) sets. Also, we studied the relationship between them and temporal open (closed) sets in the temporal topological space. We also created some properties that are achieved in temporal topological space, but are not achieved in topological space or vice versa. These new concepts in TTS open horizons for researchers to study new concepts such as continuity, separation axioms, etc.

REFERENCES

Al-Hamido, R. K. (2018). Neutrosophic Crisp Bi-Topological Spaces. *Neutrosophic Sets and Systems*, 21, 66–73.

Al-Hamido, R. K. (2019). *A study of multi-Topological Spaces* [PhD Thesis, Al-Baath University].

Al-Hamido, R. K. (2023). Infra Bi-topological Spaces. *Prospects for Applied Mathematics and Data Analysis, 1*, 8-16.

Al-Odhari. (2015). On infra topological space. *International Journal of Mathematical Archive*, 11(5), 179–184.

Al-Shami, T. M. (2017a). On supra semi open sets and some applications on topological spaces. *Journal of Advanced Studies in Topology*, 8(2), 144–153. 10.20454/jast.2017.1335

Al-Shami, T. M. (2017b). Utilizing supra α-open sets to generate new types of supra compact and supra lindel of spaces, Facta Universitatis, Series. *Mathematics and Informatics*, 32, 151–162.

Al-Shami, T. M. (2018). Supra semi-compactness via supra topological spaces. *Journal of Taibah University for Science*, 12(3), 338–343. 10.1080/16583655.2018.1469835

Al-Shami, T. M., Asaad, B. A., & El-Gayar, M. A. (2020). *Various types of supra pre-compact and supra pre-Lindel¨of spaces* (Vol. 32). Missouri Journal of Mathematical Science.

Jayaparthasarathy, G., Little Flower, V. F., & Arockia Dasan, M. (2023). Neutrosophic Supra Topological Applications in Data Mining Process. *Neutrosophic Sets and Systems, 27*, 80-97.

Mashhour, S., Allam, A. A., Mahmoud, F. S., & Khedr, F. H. (1983). On Supra topological spaces. *Indian Journal of Pure and Applied Mathematics*, 14(4), 502–510.

Mhemdia, T. M. (2024). Al-shami, Introduction to temporal topology. *J. Math. Computer Sci.*, 34(3), 205–217. 10.22436/jmcs.034.03.01

Ozturk, T. Y., & Ozkan, A. (2019). Neutrosophic Bi-topological Spaces. *Neutrosophic Sets and Systems*, 30, 88–97.

Chapter 11
Utilizing Graphics Processing Units (GPUs) for Numerical Computations

Alnoman Mundher Tayyeh
Mangalore University, India

Akram H. Shather
Sulaimani Polytechnic University, Iraq

Husam Abdulhameed Hussein
University of Samarra, Iraq

Luma Saad Abdalbaqi
Tikrit University, Iraq

ABSTRACT

Initially designed for graphics processing, GPUs (graphics processing unit) only contained fixed rendering functions. As they are parallel processors with high computational power for arithmetic calculations, GPUs have been evolving rapidly with the inclusion of programmable pipelines. The trend is for the GPU to become a generic processor (GPGPU). Furthermore, the integration between GPU and CPU cores within a single chip also tends to become a standard among processors. GPUs are also becoming the target of massively parallel systems such as supercomputers. Today, the fastest supercomputer in the world has thousands of GPUs, which has significantly increased its performance compared to its predecessors.

DOI: 10.4018/979-8-3693-3964-0.ch011

INTRODUCTION

Parallel computing aims to speed up an application, that is, reduce processing time. Therefore, the reason that led to the search for a standard for the use of parallel computing has been that, to stop problems that require a large amount of computing capacity and time. For example, a particle collision simulation problem, using a sequential algorithm and using a single processor, could take years to complete. Initially, when there was a computer with a single processor, parallelism was applied by creating a network of computers within which one could work on the same problem. That is, each computer connected to this network would have the task of working on a portion of the problem. Thus, minimizing the computing time for solving the problem [1].

Different memory architectures have been designed for parallel computing. Among these architectures we have the uniform memory access architecture (UMA), non-uniform memory access architecture (NUMA). Within data level parallelism there are four basic techniques which are: single instruction, multiple data (SIMD). Multiple instruction, multiple data (MIMD), single instruction, one data (SISD) and multiple instruction, one data (MISD). The most used is the single instruction, multiple data [2].

Over the past two decades servers, personal computers, and PMDs that use both CPUs and GPUs have become more common in the hardware industry. The appearance of these computers with heterogeneous processing technologies has promoted a new parallel programming model: heterogeneous programming model. In this environment, the programming model attempts to take full advantage of the system's available computing resources through the use of APIs and functional tools without the need to address specific hardware paradigms (for CPU or GPU) or the computational performance limitations between they. In this type of models, the calculations are managed by a host processor that provides control over other computing equipment (CPU/GPU). Parallel programming is carried out using kernel programs which implement the functionality that will be executed by the devices (computing equipment) [3].

OpenCL is an API with its own programming language called OpenCL C. It was created for hardware manufacturers by Apple and developed together with INTEL, IBM, NVIDIA and AMD. In 2008 it passed to the Kronos group to become an open standard [4]. At that time Intel and NVIDIA retired and implemented their own development environments. Its specification is based on C and C++. It has been implemented for heterogeneous hardware: CPU, GPU, DSP, FPGA, ASIC. Like other frameworks, it allows you to take advantage of the enormous potential of parallel computing and, as an advantage, it offers it openly and free of charge.

However, because it is implemented for different types of equipment, its interface and configuration becomes more complex than those that use a single type of device [5].

Each core has one processing thread at a time, with a set of registers containing the state of the thread, an ALU (Arithmetic Logic Unit) dedicated to the current thread and a large unit dedicated to task management and scheduling.

The GPU (Graphics Processing Unit) is a processing unit specialized in rendering 3D graphics. It is used on PCs, video games and currently, also on mobile devices such as the Apple iPhone. Initially, GPUs had the objective of processing only graphics and floating-point processing, which justifies the fact that they are very efficient processors in manipulating graphics and complex mathematical calculations. This efficiency is due to its ease of processing vectors or matrices as they have more processing circuits than CPUs, which in turn have more cache and flow control. In short, it is the piece of silicon that both AMD, NVIDIA or Intel manufacture and where the transistors are recorded, to put it in a common way, it is the so-called graphics card chip. A GPU is based on a silicon chip that houses a series of millions of transistors under a specific architecture and with dedicated features for each specific model. The importance of the concept itself comes to us from the APUs, for example where they also house a GPU, only they do so on a much smaller substrate and with the help of a CPU.

Nvidia CUDA Cores are parallel processors that are responsible for processing all the data that enters and leaves the GPU, performing graphic calculations whose results are seen by the end user. They are located inside the GPU as if it were a CPU and their daily tasks are rendering 3D objects, drawing models, understanding and resolving the lighting and shading of a scene, etc. This parallel computing allows the cores to work together to complete the same task. In this way, they help the CPU a lot when handling data because they help it in this task or several similar tasks at the same time. Therefore, just as we find cores inside CPUs, we find them in graphics cards, although in the latter case they are much smaller and more numerous than those we see in an AMD or Intel processor. The difference between a CUDA core and a CPU core is that the former is not as sophisticated. And graphics processing requires the need to perform many complex calculations at the same time, which is why graphics cards equip many CUDA/EUs/ cores. Stream Processors. We say the latter because AMD Radeon calls its graphics cores Stream Processors.

A striking feature of a GPU is the ability to process sections of code in parallel in a very efficient way, because, while CPUs dedicate a large amount of their circuits to control, a GPU focuses more on arithmetic units. Such parallel processing power makes the execution of complex algorithms more efficient, especially when there is a lot of data to be processed.

The continuous evolution of GPUs is mainly motivated by the need for features such as visualization of complex graphics, processing of large quantities of numerical calculations and programming flexibility. has provoked a discussion and increasingly encouraged the complete replacement of traditional CPUs by GPGPUs, that is, general-purpose graphics processing units.

Currently, technologies still do not have guaranteed efficiency when considering generic processing. Issues such as high energy expenditure and difficulty in programming software for these processors are challenges to be faced. Large companies involved in the development of these technologies (NVIDIA, Intel and AMD) compete to impose their standards on the market. As a result, investment is increasing, generating benefits for both IT professionals and simple users.

Furthermore, currently the major focus of companies developing these technologies is on the integration and collaboration between CPU and GPU, in order to take advantage of all the processing power of GPUs. GPUs are also becoming the target of massively parallel systems such as supercomputers. Today, the world's fastest supercomputer has thousands of GPUs, which has significantly increased its performance compared to its predecessors.

The use of graphics processors for general-purpose applications is not a recent approach, but it has become widespread rapidly with the C programming language-based CUDA (Compute Unified Device Architecture) architecture developed by NVIDIA in 2007. CUDA architecture allows software developers to create general-purpose programs without knowledge of graphics processors. GPU-based applications are used not only in the scientific field, but also in other high-performance fields such as image and video processing and fluid dynamics simulation [6].

Evolution of the GPU

Between the 1970s and 1980s, the first GPUS appeared with very basic fixed graphic functions, such as drawing characters, lines and arcs. From 1990 onwards there was a great advance in graphic processing with the evolution of GPUs that started to contain graphical pipelines, called shaders, such as image projection, combination textures and 3D rendering. The evolution of this hardware led to the inclusion of new programmable pipelines and support for simple precision floating point, at the beginning of the 2000s [7].

With the significant advance in their performance, GPUs became a major attraction in the area of processors, supporting double-precision floating point and several lines of code, then emerging. o the concept of GPGPU (General-Purpose Computing on Graphics Processing Units).

Currently, the main advantage of using GPGPU is having highly parallelizable thread processors, which favors the execution of several tasks at the same time. Another advantage is its high-performance hardware with an excellent cost-benefit ratio. However, developing software capable of obtaining a considerable gain for execution on this hardware is a challenge as it must respect some restrictions such as having nested and parallelizable repetition structures, high computability, regular data access and data isolation between CPU and GPU.

GPU AND PARALLEL PROCESSING

Multicore and many-core are the two main processes that have been developed since 2003 to produce microprocessors. By using multiple cores at once, a multicore processor seeks to maintain the performance of sequential programs. They started out with two cores, and with every generation after that, that number has increased. For example, the Intel Core i7 microprocessor has eight processing cores, each of which can process multiple instructions and is compatible with the entire x86 instruction set. A many-core microprocessor, on the other hand, places more emphasis on how efficiently parallel applications can run at the same time. With each new generation, the number of their tiny nuclei doubled from where they started. The 240 cores of the NVIDIA GeForce GTX 280 GPU operate as many threads connected to a single instruction in an orderly fashion. Seven additional cores share control, instruction, and memory cache with these cores [7].

Since then, as Figure 2 illustrates, GPUs have led the performance competition in floating point processing. While GPU performance has continued to improve, CPU performance improvement has slowed. As of 2009, the floating-point processing throughput ratio of GPUs to CPUs is roughly 10 to 1 (1 teraflops, or 1000 gigaflops) versus 100 gigaflops), with respect to the speed at which these chips can support resource execution options. Kirk and Hwu (2010) are the source. Different design philosophies are the cause of the performance difference between CPUs and GPUs, as Figure 3 illustrates.

Figure 1. NVIDIA GPU technology development

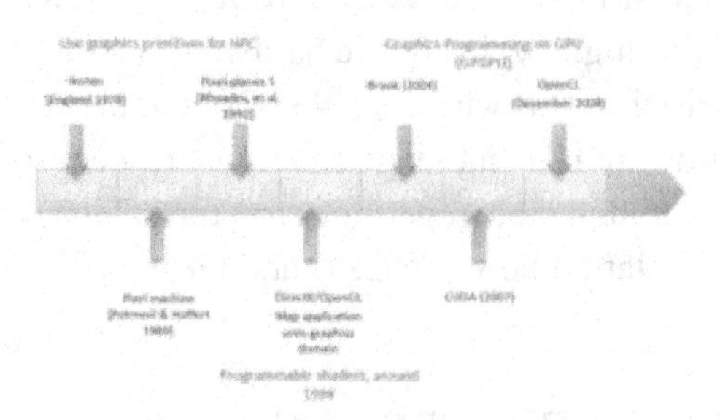

Figure 2. Performance difference between GPUs and CPUs [8]

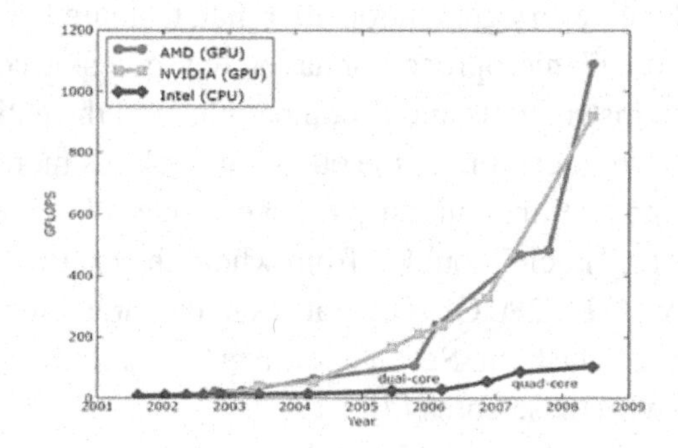

Many developers have chosen to run the compute-intensive parts of their software on GPUs due to the large performance difference between parallel and sequential execution. This difference between the performance of GPUs and CPUs exists due to the difference between design philosophies, as shown in Figure 3.

Figure 3. Comparison between the design of a CPU with four cores and a GPU

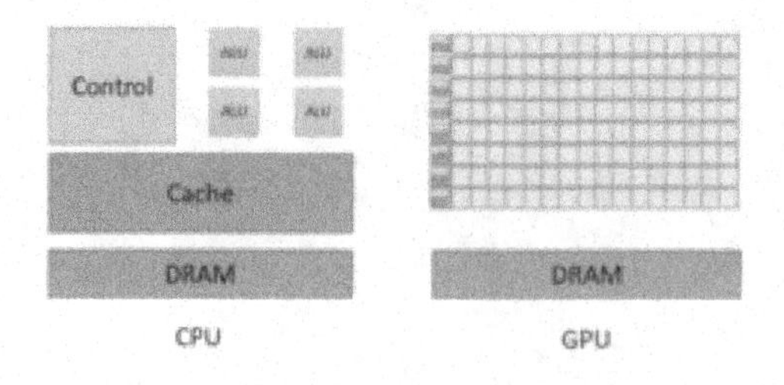

CPUs were designed with the aim of having high performance in sequential codes. However, it uses control logic that allows the execution of instructions from a single thread in parallel or even outside of their sequential order while maintaining the appearance of sequential execution. Furthermore, large caches try to reduce instruction and data access latencies. Despite this, performing calculations quickly is not guaranteed. Since 2009, the increase in the number of cores in CPUs has been exploited to improve the performance of sequential codes [7].

This made GPU vendors opt for a design that optimized the throughput of executing a large number of threads, in order to maximize the chip area and energy expenditure dedicated to calculations. floating point.

A hardware characteristic of a GPU is that due to the many threads, while some wait for access to memory that has high latency, the other threads are used to look for tasks to be executed. This feature minimizes the control logic required for each thread. Furthermore, the small cache memories are designed to help control the bandwidth demanded by multi-threaded applications by minimizing access to kirk [7].

It is easy to see that GPUs are designed to perform numerical computation and that CPUs can perform better in some tasks that GPUs were not designed for. Therefore, most applications will use CPU for the sequential parts and GPU for the numerically intensive parts. However, some programming models, such as CUDA (Compute Unified Device Architecture), are designed to support the execution of an application jointly on the CPU and GPU.

Other factors, in addition to performance, must be considered by the developer when choosing a processor. Considering the need to perform numerical computations, one of these factors is the support for the IEEE (Institute of Electrical and Electronics Engineers) floating point standard. Since the introduction of the GeForce GTX 8800 or simply G80, in 2006, GPUs have supported the IEEE standard, although this was not common in the first generations. Processors from different vendors supporting this standard provide predictable results. Furthermore, newer GPUs have faster approaches to double-precision floating point execution.

Another factor is the popularity of the processor, mainly due to the cost of developing an application on a parallel computing system. However, with the popularity of GPU in the PC market, massively parallel computing was characterized as a popular product, which made GPUs economically attractive to application developers.

When it is possible to develop an application suitable for parallel execution, a GPU implementation can achieve a speed approximately 100 times greater than a sequential execution . There are several levels of parallelism detail that can be achieved with the CUDA architecture, which has a programming model to facilitate parallel implementation and data sending management [7].

In addition to the countless current computing applications, many interesting applications of the future can enjoy the benefits of high performance from parallel computing such as detailed simulations in the medical and biological areas; video and audio coding and manipulation, image synthesis and high-resolution display on high-definition televisions (HDTV); manipulation of user interfaces on touch-sensitive screens of cell phones and video monitors; and mainly, applications in the gaming market, major investors in the development of GPUs.

GPGPU

GPU hardware designs are becoming more similar to high-performance parallel computers due to the shift towards unified processors. Researchers observed the improved performance of GPUs and started investigating their application in solving complex engineering and scientific problems. GPUs were originally designed specifically for functions needed by graphics application programming interfaces (APIs). In order to utilize computational resources, a programmer needed to employ native graphical operations using OpenGL or DirectX functions as outlined by Kirk and Hwu in 2010.

The beginning of the use of GPUs for programming general purpose applications boosted the development of facilities in the architecture of GPUs and mainly, in programming with the introduction of the CUDA architecture (Compute Unified Device Architecture) presented by NVIDIA in 2006 [NVIDIA b]. The CUDA architecture is a programming model and a parallel computing platform that aims to improve the use of the GPU's processing power. The CUDA programming model provides programmers with simple abstractions of the hierarchical organization of threads, synchronization and memory allowing the appropriate implementation of programs for the GPU. CUDA supports programming languages such as C, C++, Fortran, OpenCL and DirectX [17].

NVIDIA's G80 GPU was the first generation of GPUs prepared for CUDA and support for the C language, which has increasingly improved in the following generations: G200, launched in 2008; Fermi, launched in 2009 and; Kepler, launched in 2012.

GPU ARCHITECTURE

Multi-threaded SMs (Streaming Multiprocessors) are the foundation of NVIDIA GPUs that use CUDA. Figure 4 depicts a building block consisting of two SMs; the exact number of SMs varies from generation to generation. Each SM is made up of

Streaming Processors (SPs), which share a cache of instructions and control logic. Each GPU supports a maximum of 4 GB of graphics data rate (GDDR) DRAM. Unlike the system DRAMs found on the motherboard of the central processing unit (CPU), these GDDR DRAMs are primarily used as graphics memory. They are used for 3D rendering in graphics applications and as off-chip memory in computing; their latency is slightly higher than that of a conventional memory system, but their bandwidth is extremely high. The G80 GPU introduced the CUDA architecture, which has 86.4 GB/s of memory bandwidth and 8 GB/s of communication bandwidth with the CPU, which should increase as bandwidth grows. RAM and CPU bus speed were both increased [7].

With 128 SPs (16 SMs, each with 8 SPs) and 500 GBaflops of internal memory, the G80 GPU is extremely parallel. Every SP contains both a multiplication unit and an addition unit (MAD). In addition to SQRT and other floating-point operations, other special function units can be used. Each SP is capable of running thousands of threads per application. The most recent GT200 has 240 SPs, or more than 1 teraflop, and can support up to 30,000 threads on the chip, with 1024 threads per SM. Intel CPUs typically support 2-4 threads per core, which is less than simultaneous multithreading (7).

Figure 4. Architecture of a GPU with CUDA support [7]

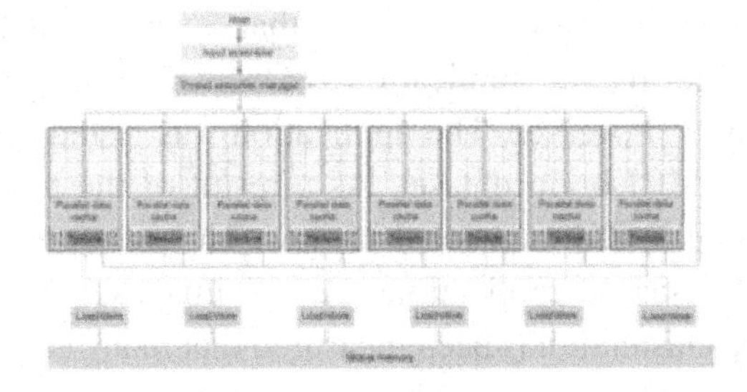

Image Processing and Level Set Method on GPU

Before the CUDA C language was developed, using the graphics card for general purpose programs required a very difficult process. In addition, as libraries such as OpenGL and DirectX used in GPU programming are difficult to use, efforts have been initiated to develop programming models that everyone can benefit from. As a result of the studies, some structures (such as BrookGPU, CTM) have been developed.

With these structures, attempts were made to run general-purpose programs on the GPU. The BrookGPU project also marked the beginning of the CUDA C language.

CUDA was developed by [17] and the first SDK (Software Development Kit) was published in February 2007. Since this date, CUDA has been used in many branches of science for both commercial and research purposes [10]. For example, in the field of data mining, in 2013, You Li et al. [11] ran the k-Means algorithm, which plays an important role in cluster analysis, on GPU. The study is divided into two: low-dimensional dataset and high-dimensional dataset. Registers were used for low-dimensional datasets, and shared memory was used for high-dimensional datasets. Analyzes have shown that this approach is faster than previous GPU-based k-means algorithms.

The first GPU implementation of the level set algorithm was made by Rumpf and Strzodka [12] in 2001. Based on the idea of benefiting from the high performance and high bandwidth of modern graphics cards for numerical calculations, the numerical operations of the level set algorithm were carried out on the GPU. In the study, 2-dimensional (1282) images were used and the calculation time of a single step in the first-order base level cluster model was measured as 2ms. In addition, it was stated in the study that 3-dimensional studies can be done with the development of graphics hardware.

In the study conducted by Aaron Lefohn et al. [14] in 2002, 2-dimensional and 3-dimensional level set solution was implemented on GPU basis. The developed application is the first GPU-based application that can calculate the curvature flow as well as the speed term. The application is OpenGL based and ran on the ATI Radeon 8500 graphics processor. The application was tested on 256x256 images and a 256x256x175 MRI brain dataset. As a result of the tests, it takes 4ms to calculate a single step in 2-dimensional images. This is approximately similar to highly optimized CPU-based applications. For 3D images, although the developed application is 2 times faster, it has been measured that many calculations are accelerated up to 10 times. It has been emphasized that speedups of up to 20 times can be achieved with some improvements. In 2004, Aaron Lefohn et al. In another study conducted by [14], the study conducted in 2002 [13] was improved. The solution proposed in the study is to transfer the narrow-band algorithm in real time. With the new algorithm, the data of the level set curve is loaded into the 2D texture memory through the multidimensional virtual memory system. When the curve moves, this texture-based representation is dynamically updated by the GPU-to-CPU message passing algorithm. Thus, the user can follow the curve development while the level set algorithm is running. The developed application was run on an Intel Xeon 1.7 GHz processor, ATI Radeon 9800 Pro GPU with 1 GB RAM. The level set algorithm was run on 256x256x175 sized data and 70 steps per second were completed in tumor segmentation. In contrast, the optimized CPU version (Insight Toolkit 2003)

[15] was able to complete 7 steps per second. According to the results obtained in the tests, it provided a 10 to 15 times speed increase compared to the CPU-based implementation (Insight Toolkit 2003).

In 2013, Andrei C. Jalba et al. [16] sportified level set method was used for 3-dimensional surface creation and was implemented on GPU. In this study, a speed increase of up to 20x was achieved compared to CPU-based applications on the subject. On the GPU side, it was compared with the work done by [16]. In the comparisons made, speed increases of up to 5x were achieved despite the narrow band being chosen larger.

Fermi GPU Architecture

The Fermi architecture, introduced in April 2010, is depicted in Figures 5 and 7. This architecture introduced new instructions for C++ programs, including dynamic object allocation and exception handling support. Considerations. Each Streaming Multiprocessor (SM) contains 32 CUDA cores, resulting in a combined total of 512 cores. Each clock cycle allows for a maximum of 16 double precision operations per Streaming Multiprocessor (SM). Additionally, each Streaming Multiprocessor (SM) contains 16 load and store units, enabling the calculation of destination source addresses for 16 threads per clock cycle. It also includes 4 Special Function Units (SFUs) that handle transcendental instructions like sine, cosine, and square root. Each SFU is independent from the issuing unit, enabling it to issue a new instruction even when the SFU is occupied [17].

The Fermi GPU features 6 high-speed GDDR5 DRAM interfaces, each with a width of 64 bits. The 40 address bits can accommodate an address space of up to 1 terabyte for DRAM GPU, as stated by Kirk and Hwu in 2010.

Figure 5. Fermi GPU architecture

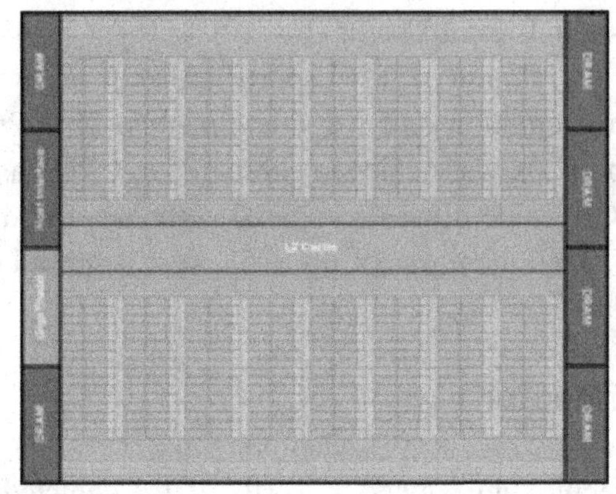

(Nickolls and Dally, 2010)

Figure 6. Architecture of a fermi GPU microprocessor

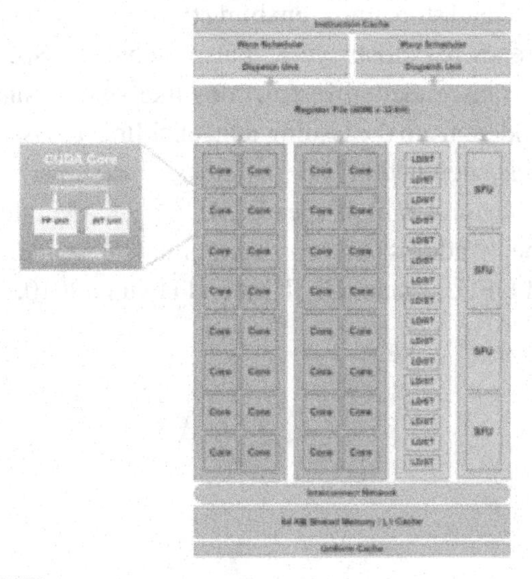

(Nickolls and Dally, 2010)

UNIFIED ARCHITECTURE

A number of processor manufacturers are opting to integrate GPUs and CPUs onto a single chip, a process known as unified architecture. Despite advancements in processor performance and energy efficiency, the fundamental functions of the CPU core and GPU core continue to be distinct. The lack of collaboration among them in the execution of a particular program leads to diminished efficiency.

In 2006 there was a merger between ATI and AMD, which in the same year announced the beginning of the development of a technology integrating GPU and CPU, making the announcement official in 2010 with the launch of the AMD Fusion processor . Soon after, in 2011, Intel released the new Sandy Bridge microarchitecture, with the same purpose of integration with NVIDIA's GPU.

Figure 7. Internal organization of AMD fusion

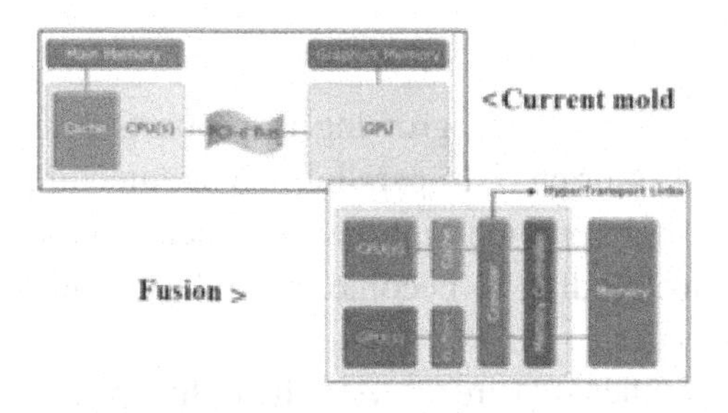

A new technique for unified architecture is presented by Yang et al. (2012). According to tests presented by Yang et al. (2012), this technique improves computer performance by approximately 20% compared to the same cores working in separate functions.

The interconnection between graphics processors and main processors does not require any new hardware technology and consists of making the GPUs collaborate with the CPUs, allowing the GPU cores to perform the functions calculation, while the CPUs load and prepare the data from the memory that the GPUS will need. As GPUs process data faster, CPUs are in charge of discovering what the GPU will need next, fetching this data from memory and delivering it ready-made. This eliminates the need for the GPU to consult memory and consequently its performance improves as it focuses on data processing [9].

USE IN MASSIVE PARALLEL SYSTEMS (MPS)

With the development of facilities in the architecture of GPUs and mainly in programming, GPUs also began to be widely used in massively parallel systems. One example is the Titan supercomputer. Launched in 2012, it is considered by Top500 to be the fastest in the world, with a processing capacity of 20 petaflops.

It is a Cray XK7 massively parallel system with 18,688 compute nodes distributed across 200 cabinets. Each node contains an AMD Opteron 6274 processor with 16 cores clocked at 2.2 GHz and 32GB of DDR3 memory. It also features an NVIDIA K20 Tesla GPU and 6GB of fast memory. The machine has a total memory capacity of 710 TB. The Titan is ten times faster and five times more energy efficient than its predecessor, the Jaguar 2.3 petaflops, while taking up no additional space [17]. This is because the Tesla K20 GPU consumes less power and costs less.

CONCLUSION

The number of cores has been growing in proportion to the growth in area availability of the processor chip. In parallel, the architecture of GPUs continues to evolve and increasingly provide high performance in processing graphics and numerical calculations, tending to become generalized with the development of new GPGPU technologies.

Many advantages of the GPU are not yet fully utilized due to the difficulties that still exist in software programming, which also tends to improve with the development of new techniques, languages, compilers and improvement of programming models.

Scalable parallel computing on GPU is a relatively new field that has resulted in the rapid development of numerous applications. This encourages processor development companies to investigate and implement new optimizations. You are becoming faster. As a result, the number of highly parallel systems based on GPUs is growing in tandem with their performance improvements.

REFERENCES

6. Al-Aaraj, H., Al Sayeh, A. M., & Mohammad, A. T. (2024). Application of approximation algorithms to the detection and categorization of diseases. *Tamjeed Journal of Healthcare Engineering and Science Technology, 2*(1), 20–28. Retrieved from https://tamjed.com/index.php/TJHEST/article/view/42

15. Aloran, A., & Al-shawagfeh, D. M. (2024). The effect of work environment on the efficiency of lungs function in people infected with Corona virus among elderly workers in the factories of the southern region in Jordan. *Tamjeed Journal of Healthcare Engineering and Science Technology, 2*(1), 13–19. .10.59785/tjhest.v2i1.31

10. Al-Rweis, A., Zakaraya, Z., Al-Omari, L., & Abdul-Aziz, K. (2024). Impact of smoking on Galectin-3 and GDF-15 among pregnant women. *Tamjeed Journal of Healthcare Engineering and Science Technology*, 2(1), 1–12. 10.59785/tjhest.v2i1.37

16. Jalba, A. C., Van Der Laan, W. J., & Roerdink, J. B. T. M. (2013). Fast Sparse Level Sets on Graphics Hardware. *Visualization and Computer Graphics*, 30-44.

7. Kirk, D. B., & Hwu, W.-W. (2010). *Programming Massively Parallel Processors: A Hands-on Approach* (1st ed.). Morgan Kaufmann Publishers Inc.

14. Lefohn, A., Kniss, J. M., Hansen, C. D., & Whitaker, R. T. (2004). A Streaming Narrow-Band Algorithm: Interactive Computation and Visualization of Level Sets. *IEEE Transactions on Visualization and Computer Graphics*, 10(4), 422–433. 10.1109/TVCG.2004.218579970

13. Lefohn, A., & Whitaker, R. (2002). *A GPU-Based Three-Dimensional Level Sey Solver with Curvature Flow, Scientific Computing and Imaging Institute Technical Report, UUCS-02-017*. University of Utah.

11. Li Y., Zhao K., Chu X. & Liu J. (2013). Speeding up k-Means Algorithm by GPUs. *Journal of Computer and System Science, 79,* 216-229.

8. Mann, D. (2010). GPU Accelerated Novelty Detection In Long-Term Audio Files. 10.13140/RG.2.1.1059.6880

1. Mantas, José & Asunción, Marc & Castro, Manuel. (2016). *An Introduction to GPU Computing for Numerical Simulation*. .10.1007/978-3-319-32146-2_5

2. Navarro, C., Hitschfeld, N., & Mateu, L. (2013). A Survey on Parallel Computing and its Applications in Data-Parallel Problems Using GPU Architectures. *Communications in Computational Physics*, 15(2), 285–329. 10.4208/cicp.110113.010813a

5. Peng, Z., Gong, Q., Duan, Y., & Wang, Y. (2017). The Research of the Parallel Computing Development from the Angle of Cloud Computing. *Journal of Physics: Conference Series*, 910, 012002. 10.1088/1742-6596/910/1/012002

12. Rumpf, M., & Strzodka, R. (2001). Level Set Segmentation in Graphics Hardware. *IEEE International Conference on Image Processing (ICIP'01)*, Thessaloniki, Greece.

3. Saidu, C., Obiniyi, A., & Ogedebe, P. (2015). Overview of Trends Leading to Parallel Computing and Parallel Programming. *British Journal of Mathematics & Computer Science.*, 7(1), 40–57. 10.9734/BJMCS/2015/14743

17. Sinha, S. N., Frahm, J. M., Pollefeys, M., & NVIDIA. (2006). GPU-based video feature tracking and matching. In *EDGE, workshop on edge computing using new commodity architectures* (*Vol. 278*, p. 4321).

4. Stone, J. E., Gohara, D., & Shi, G. (2010). OpenCL: A Parallel Programming Standard for Heterogeneous Computing Systems. *Computing in Science & Engineering*, 12(3), 66–72. 10.1109/MCSE.2010.6921037981

9. Yang, Y., Xiang, P., Mantor, M., & Zhou, H. (2012). Cpu-assisted GPGPU on fused CPU-GPU architectures. In *Proceedings of the 2012 IEEE 18th International Sympo- sium on High-Performance Computer Architecture*. IEEE Computer Society. 10.1109/HPCA.2012.6168948

Chapter 12 Validation and Verification of Numerical Models

Bahaa Hayder Mohammed
University of Samarra, Iraq

Yas Khudhair Abbas
University of Samarra, Iraq

Basava Ramanjaneyulu Gudivaka
Raas Infotek, USA

Sri Harsha Grandhi
Intel, Folsom, USA

ABSTRACT

The conceptual differences between the two expressions are extremely important for the development of this work, since the code used, although analysed through example cases, does not have validation like the proposal of the present work. For a more in-depth discussion on the definitions of validation and verification suggested. This work's main objective is to validate a code based on the finite element method used to simulate incompressible fluid flows. Numerical simulations were carried out for the cylindrical specimen. The two conditions for which data were found in the literature were simulated and experimental tests were carried out. From the simulation results, pressure coefficient graphs around the cylinder were obtained. The curves resulting from the simulations did not coincide with those obtained through experimental tests, for the reason previously explained. However, the pressure coefficient curves resulting from the simulations fit reasonably well with those

DOI: 10.4018/979-8-3693-3964-0.ch012

originating from literature data.

INTRODUCTION

Fluid dynamics studies the effects caused by the interaction of the fluid with the environment. Properties like the Reynolds number, which measures the ratio of inertial forces to viscous forces, define flow categories in fluid dynamics, such as laminar, turbulent, and transitional flows. Surface roughness, characteristic length, fluid velocity, and viscosity are important variables that affect these categories. The flow categories are defined according to the shape of the immersed body, the speed and orientation of the flow, as well as the behavior of the properties inherent to the fluid, including viscosity and density. Therefore, in order to model a problem, it is necessary to define which characteristics of the fluid and flow regime it is dealing with. When modeling fluid dynamics problems, defining fluid properties (such as viscosity, density, and flow regimes, such as laminar and turbulent) is essential because it affects the choice of suitable equations, boundary conditions, and numerical techniques, all of which have an impact on the precision and dependability of simulations. One must then define the properties of the fluid at rest and in motion (Wang, A. X., et. al, 2022).

The essence of the study of fluid flow is an integration between theory and experiment. Fluid theory has a large set of basic laws that govern the behavior of phenomena and that allows, to a certain extent, the definition of analytical solutions to the proposed problems (Tamrakar, A., et. al, 2018). "The traditional fluid model used in physics is based on a set of partial differential equations known as the Navier-Stokes equations" (Heidari, E., et. al, 2022). Although the equations have been known for a long time, currently "the mathematical theory of this class of equations is not sufficiently developed to allow obtaining analytical solutions in arbitrary regions and general boundary conditions" (Chattouh, A., et. al, 2022). Through the use of discretized equations and potent computational techniques, numerical modeling overcomes the limitations of analytical solutions by enabling the study of difficult, real-world fluid flow problems that are unsolvable analytically. This allows for the thorough analysis and optimization of fluid systems. This is due to its non-linearity, appearance of turbulent regime and occurrence of instabilities. The appearance of turbulent regimes and instabilities greatly exacerbates problems with fluid flow because turbulence produces chaotic and unpredictable fluctuations that are hard to simulate and demand a lot of processing power. This makes it difficult to find exact solutions and calls for sophisticated modeling methods. The exact solutions that govern the phenomena serve mainly for idealized situations, and cannot be extended to real situations, due to the demand for exaggerated simplifications. Therefore, there is still much to be done (Bazant, M. Z., et. al, 2016). A more intensive study of numerical modeling involves, among other things, carrying out experiments in order to obtain data so that the code can be validated.

The numerical model based on incompressible fluid flow has several formulations. Because of the nonlinear nature of the Navier-Stokes equations and the difficulties in modeling turbulence and multiphase flows, even the most sophisticated mathematical theory is often unable to accurately predict fluid behavior in complex or heterogeneous environments, requiring reliance on empirical data and sophisticated computational techniques. One of them, the mixed model, also known as velocity-pressure, was used to implement the code. According to (Reddy, J. N., et. al, 2005), the mixed model consists of the natural and direct formulation of the differential equations that make up the system.

The development of numerical tools has become of fundamental importance in the study of fluid dynamics. Computational fluid dynamics (CFD) approaches are improving rapidly and have made great progress in recent decades, allowing the simulation of increasingly complex phenomena that were previously possible only through obtaining experimental data, such as those obtained through wind tunnels. By combining intricate physics, high-resolution meshes, and sophisticated algorithms, advances in computational fluid dynamics (CFD) allow for more accurate simulations. This reduces the discrepancy between theoretical and experimental predictions, improving our comprehension and ability to predict fluid behavior.

In computational fluid dynamics, specifically the numerical simulation of internal aerodynamics, the focus of this work, can be modeled and solved through several existing physical-mathematical models, known to be discrete. By discretizing the continuous fluid domain into a finite set of elements or volumes, discretization in numerical methods enables the complex equations governing fluid flow to be solved numerically, enabling detailed and accurate simulations of internal aerodynamic problems. The numerical code used in this work uses one of the possible discretization methods, the finite element method, which has already been analyzed in terms of comparing its results with classic literature cases (Bhalekar, S., et. al, 2016).

However, in order to complement the analysis of results obtained through codes intended for the numerical solution of equations for problems involving flow phenomena, it is of interest to carry out experimental procedures, thus allowing the validation of the discrete model.

Therefore, the main objective of this work is the validation of a code based on the finite element method, with reference to experimental data obtained from carrying out a series of experiments in the wind tunnel. Enabling the completion of this work, the use of a high-precision numerical tool capable of solving problems of engineering interest.

METHODOLOGY

(1) Experimental methodology for pressure taking

Script developed to assist a future generation of meshes. As it is a profile without an algebraic equation, the Clark-Y profile was created using SpLine on top of a perpendicular image of the specimen. The Clark-Y profile lacks an algebraic equation, which makes mesh generation more difficult. Instead, the profile must be approximated numerically or using CAD software, requiring sophisticated scripting and interpolation techniques to guarantee accurate representation and seamless mesh transitions. Figure 1 illustrates the creation of the curve for the case in which the angle of attack β is zero. The coordinates were obtained for the profile positioned in a test section 300 mm long and 150 mm high, with the profile centered at $x_o = 100\ mm$ and $y_o = 75\ mm$. The points that generate the profile curve are in Table 1.

Figure 1. Clark Y profile for zero angle of attack—Dimensions in millimeters

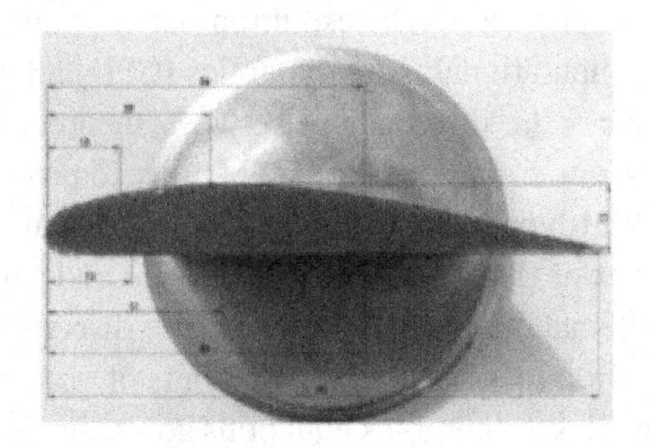

Table 1. Clark-Y Profile Points for $\beta = 0^o$

Point	x [mm]	y [mm]	Ponto	x [mm]	y [mm]
Point 1	149.2695	72.3013	Point 13	65.1450	82.8669
Point 2	149.9850	73.2045	Point 14	59.8545	81.5532
Point 3	148.9657	73.9061	Point 15	54.8309	79.7413
Point 4	145.0228	74.5413	Point 16	52.0943	76.5978
Point 5	139.3714	75.3956	Point 17	52.0150	75.7500
Point 6	132.4509	76.9797	Point 18	52.1362	75.0800
Point 7	119.8848	80.0161	Point 19	53.6226	73.4261

continued on following page

Table 1. Continued

Point	x [mm]	y [mm]	Ponto	x [mm]	y [mm]
Point 8	108.5750	82.3058	Point 20	56.4954	72.7095
Point 9	94.5392	83.9497	Point 21	59.4215	72.4527
Point 10	81.3050	84.4213	Point 22	61.9045	72.3013
Point 11	75.4470	84.2672	Point 23	67.1650	72.3013
Point 12	69.5572	83.7157	Point 24	83.3250	72.3013
			Point 25	108.5750	72.3013

The dimensions in the sketch were to ensure that the curve passed exactly through the points where the pressure points on the profile, present in the test piece, are located. Similarly, the profiles for β equal to two, five and ten degrees are shown in Figure 2 to Figure 4. It is crucial to confirm that the curve passes through the test piece's pressure points in order to validate the experimental results, show that the measurements were accurate, and guarantee that the flow characteristics were appropriately represented in the analysis.

Figure 2. Clark Y profile for angle of attack of two—Dimensions in millimeters

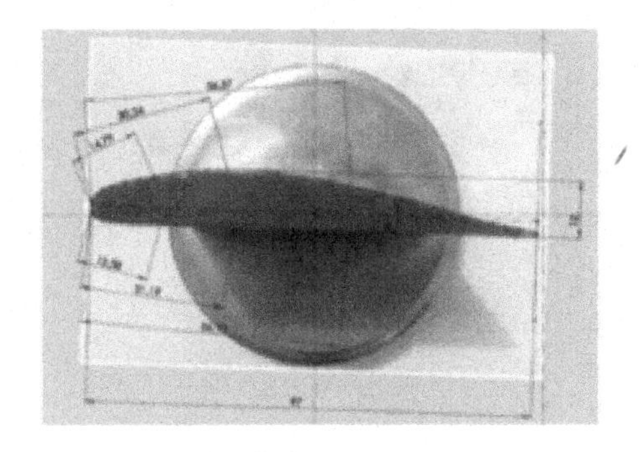

Figure 3. Clark Y profile forangle of attack of five—Dimensions in millimeters

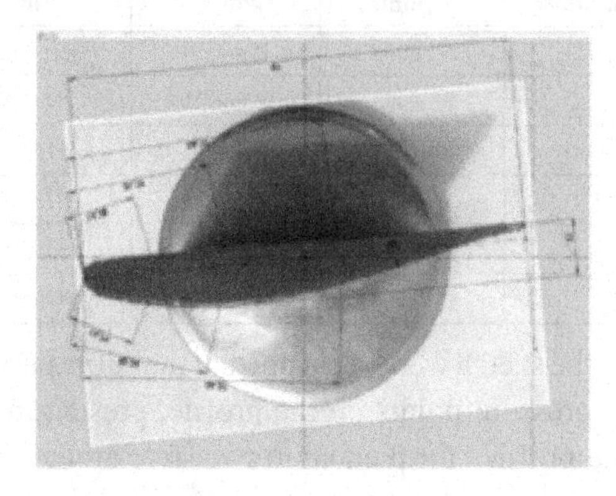

Figure 4. Clark Y profile for angle of attack of ten—Dimensions in millimeters

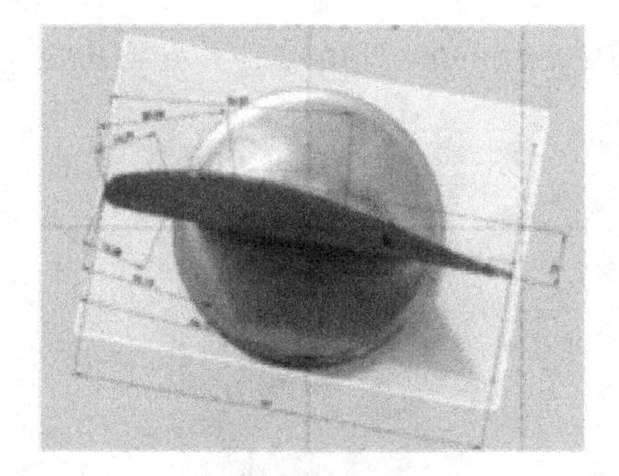

(2) Tunnel instrumentation

To make it possible to obtain all the information necessary for the work, a series of measuring instruments were used in conjunction with the tunnel.

The pressure taps were coupled to a Krell System inclined column manometer, illustrated in Figure 5, for measurements of up to 10 mmCA. For cases in which pressures were higher than this value, a U-type manometer with an end scale of 100 mmCA was used. The Krell System inclined column manometer and the U-type manometer measure differential pressure, providing precise data on pressure variations across the test section, crucial for calculating pressure coefficients and analyzing flow behavior.

Pitot tubes were used for pressure measurements. The capacities of extra and pilot tubes differ, but they both measure fluid flow. To determine velocity, a conventional Pitot tube measures stagnation and static pressure simultaneously. Additional Pitot tubes collect pressures at several locations, such as Pitot-static or multi-hole versions, and provide comprehensive flow profiles perfect for intricate analysis. The cylindrical body has one outlet, the Clark-Y has seven outlets and an extra Pitot Tube was also used to obtain the static flow pressure at a point upstream of the test section.

Figure 5. Krell system inclined column pressure gauge

To obtain the local atmospheric pressure, it was also necessary to use an Instrutemp brand Barometer, which measured a pressure of 922.2 hP to 9.222×10^4P a. The Instrutemp Barometer utilizes a calibrated aneroid capsule, responsive to atmospheric pressure changes. This capsule's movement is conveyed via a mechanical linkage to a display, featuring pressure units like inches of mercury or millibars for easy reading of current atmospheric pressure.

The temperature at which the experiments were carried out was monitored through the sensor of an Instruter Hot Wire Anemometer. The temperatures measured from the anemometer made it possible to obtain the density and dynamic viscosity values, interpolated from Appendix A of Fox, (Fox, R.W., et. al, 2014), and arranged according to Table 2.

(3) Cylinder Procedures

First, the test specimen was fixed in the test section and the frequency inverter that controls the fan rotation was turned on. Next, the hot wire anemometer was attached to the support downstream of the test section, responsible for measuring the flow temperature.

Next, the frequency in the inverter was configured, which is directly correlated with the flow speed, i.e., the Reynolds number. Tests were carried out at frequencies of 7; 7.8; 10.0; 10.4 and 13 Hz, totaling five different initial conditions.

The cylinder pressure tap is then connected to the U-tube pressure gauge and the angle α between the orifice and the flow direction is adjusted with the aid of the protractor built into the test section.

After adjusting the fan frequency in order to achieve the desired current speed U1 i.e., the desired Reynolds number, pressure readings were taken on the manometer, varying the angle α, illustrated in Figure 6, from 0 to 360, 10 each time, except for a few occasions when it was deemed necessary to take pressure every 5 minutes.

Pressure measurements were taken for different Reynolds numbers, within the laminar flow regime, by controlling the fan frequency. Finally, from the pressure values obtained, the pressure coefficients (Cp) were calculated as a function of α, and the results are presented in int coming sections.

Table 2. Temperatures measured in the tests and their dependent properties

Frequency (H)	**Temperature (C)**	**Density (kg/m³)**	**Visc. Dynamics (N · s/m2)**
7.0*	20	1.21	0.000018
7.8*	15.2	1.23	0.000018
10.0*	24	1.21	0.000018
10.4*	21.3	1.21	0.000018
13.0*	18.7	1.22	0.018100
7.0**	22.3	1.2	0.000018
8.5**	22.5	1.2	0.000018
10.0**	15	1.23	0.000018

Note: * – Frequency referring to the cylinder. ** – Frequency referring to Clark-Y.

(4) Procedures for Clark-Y

For the Clark-Y specimen, the determination of the pressure coefficients around the profile was carried out by also fixing the profile in the wind tunnel similarly to the cylinder. The profile has seven pressure tap outlets, illustrated in Figure 7.

Data were taken at Reynolds numbers within the laminar regime. The pressure coefficients were calculated for values of β and as a function of the ratio (x/L) between the position x of the point on the ordinate axis and the mean chord of the aerodynamic profile. The angle of attack β, defined as the angle formed between the middle chord of the profile and the flow direction, was set at 0, 2, 5 and 10. Tests were carried out at frequencies of 7.0; 8.5 and 10Hz. Counting three different initial conditions and four angles of attack, a total of twelve different configurations for the Clark-Y were tested.

The air properties used in the calculations are shown in Table 2. After carrying out the tests, the pressure coefficients were calculated in the positions of the profile holes and the CP are presented as a function of the ratio (x/L).

Figure 6. Angle α between the cylinder pressure take-off orifice and the flow direction

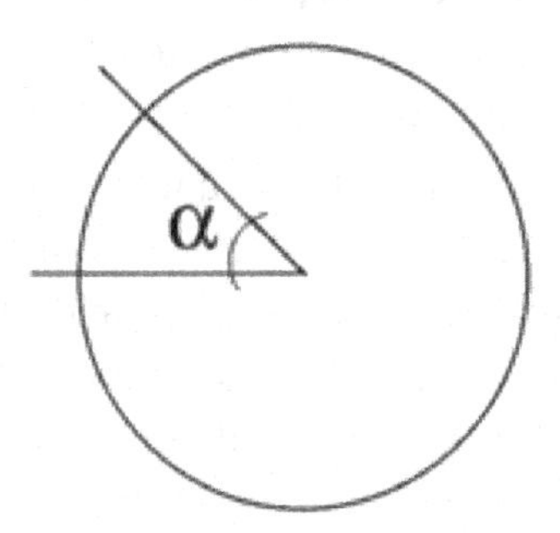

Figure 7. Outputs for the pressure gauge pressure holes of the Clark-Y

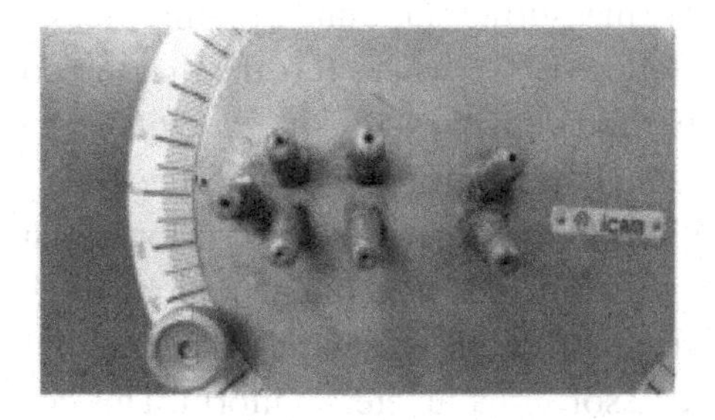

(5) Methodology for numerical simulation

The code used was implemented by (Fernandes, J. W. D. (2014). The simulations were carried out considering the discretization in two-dimensional Cartesian space, viscous incompressible fluid, in the laminar regime. The process of discretizing simulations for a viscous, incompressible fluid in two-dimensional Cartesian space entails breaking the domain up into a grid of finite elements, applying the Navier-Stokes equations at each node, and making sure the fluid parameters and boundary conditions are appropriately represented.

The laminar characteristic was desired because the code does not have a turbulence model. Without a turbulence model, numerical simulation codes fall short in accuracy and reliability for fluid flow simulations. These models are essential for capturing the chaotic nature of turbulent flows, crucial for predicting energy dissipation, mixing, and momentum transfer. This omission leads to major inaccuracies in high Reynolds number flows, affecting design and performance evaluations. Therefore, we worked below the critical Reynolds number to prevent effects due to turbulence from excessively affecting the results.

(Fernandes, J. W. D. (2014) opted to use a mixed finite element to carry out the computational implementation. This element has a triangular shape, with three nodes for pressure, located at the vertices of the triangle, and six nodes for velocity, located at the vertices and midpoints of the edges. Because of their accuracy in simulating fluid flow properties, stability in numerical simulations, and flexibility in fitting complex geometries, triangular components with particular node configurations for pressure and velocity were selected.

The literature data for the cylinder that were found are from (Lam, K., et. al, 1995), who presented the behavior of Cp for Reynolds equal to 1.9×10^4, illustrated in Figure 8 and (Igarashi, T., et. al, 1981) presented data for the cylinder at a Reynolds equal to 3.5×10^4, as shown in Figure 8. Methods for calculating pressure coefficients (Cp) involve measuring static and dynamic pressures, using the formula Cp = (P - P_ref) / (0.5 * ρ * U^2), where P is the measured pressure, P_ref is the reference pressure, ρ is fluid density, and U is velocity.

For the numerical simulations in the cylinder, the code configurations maintained the same conditions as the experimental tests for which there is literature, so that the results could be compared later. In simulations of fluid dynamics or heat transfer within cylindrical settings, it's essential to keep settings like grid resolution, boundary conditions, and solver parameters uniform. This consistency ensures that any observed differences in flow or thermal patterns are due to intentional changes in the physical or geometric variables, not setup inconsistencies. Such standardization is crucial for ensuring the accuracy and real-world relevance of these models.

Because Clark-Y is a geometry that does not have an algebraic discretization, to assist future work, the profile was digitally discretized, based on the specimen, for four different angles of attack and the procedures. The Clark-Y airfoil's complex shape, combining curves and straight lines, defies simple algebraic representation. This limitation underscores the need for numerical methods or approximations in aerodynamic analysis. Accurate CFD simulations rely on these advanced tools to capture the airfoil's aerodynamic characteristics effectively.

Figure 8. Literature data for $\mathbf{C}_p$, as a function of α, of the cylinder for $\mathbf{Re} = 1.9 \times 10^4$ & $\mathbf{Re} = 3.5 \times 10^4$

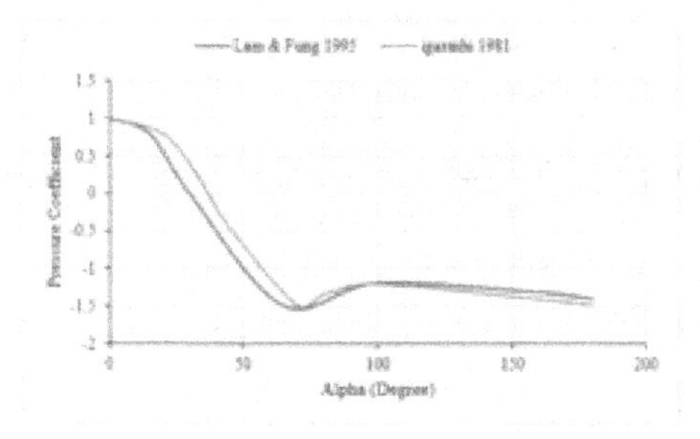

(6) Cylinder Geometry

The flow domain was chosen compatible with the dimensions of the wind tunnel test section. Using the open-source software Gmsh3, the cylinder geometry was configured as shown in Figure 9 (Geuzaine, C., et. al, 2009).

Figure 9. Geometry used to generate the cylinder meshes

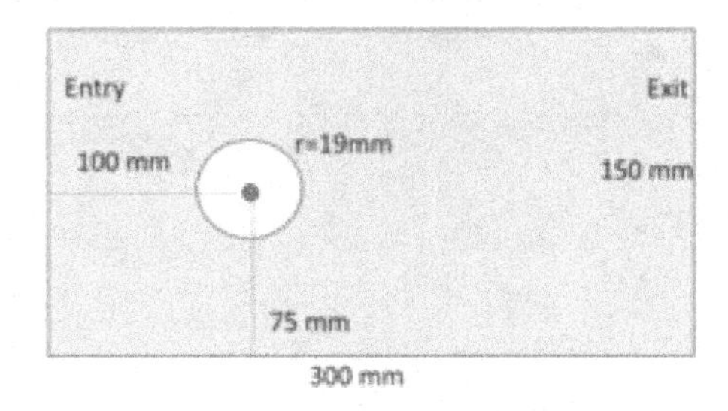

Once the geometry was defined, the parameters were chosen aiming to further refine the mesh in the region of interest, close to the cylinder, allowing elements far from this area to be larger in size. The meshes used are unstructured.

After these configurations, a mesh with 440 elements, illustrated in Figure 10, and one with 1146 elements, illustrated in Figure 11, were generated. These meshes were used to simulate the behavior around the cylinder for the two physical conditions for which there are experimental data. and in the literature, Reynolds equal to 1.9×10^4 and 3.6×10^4, respectively.

(7) Validation methodology

The six-cylinder pressure profiles were compared: those measured experimentally, those obtained through simulations and those found in the literature, through graphs of pressure coefficient (C_p) by α, as these are the ones that best allow analyzing the three cases of dimensionless way.

Figure 10. Mesh generated with 440 elements for the cylinder simulations

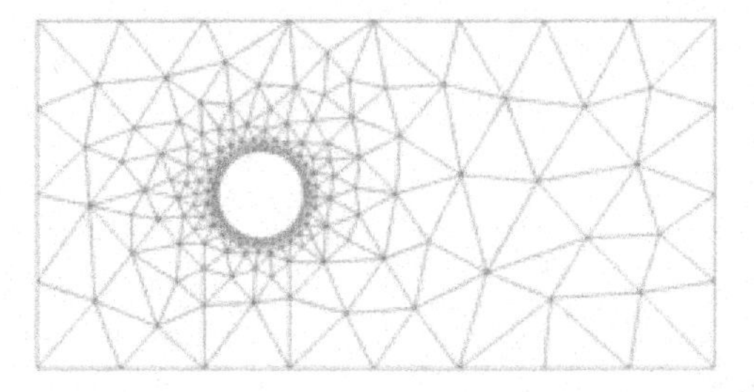

Figure 11. Mesh generated with 1146 elements for the cylinder simulations

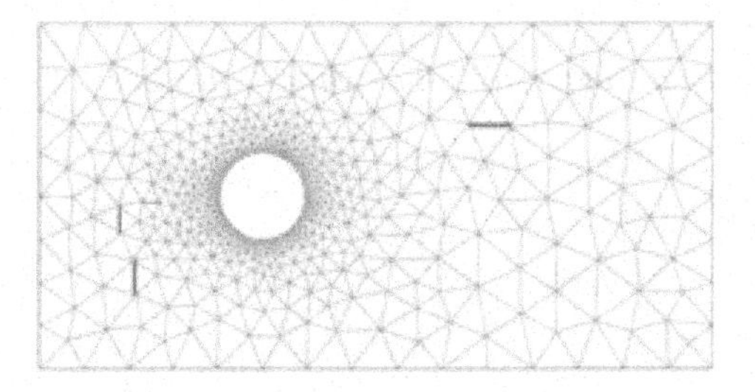

RESULTS AND DISCUSSION

This section presents the results obtained through experimental procedures and numerical simulations. Internal flow was analyzed, for low flow velocities, in a laminar regime. After the analyses, the data relating to the cylinder were compared and a validation of the code through analysis of the results was carried out.

(1) Wind Tunnel Results

The results obtained from the experimental tests carried out in the wind tunnel of the mechanics laboratory were the flow configurations at the inlet and the pressure profiles around the test specimens. In order to study the flow configurations and pressure distributions around various aerodynamic profiles, the mechanics laboratory wind tunnel was equipped with sophisticated instrumentation for measuring pressure and velocity profiles, high-resolution data acquisition systems, and precise control over flow conditions. Two instrumented bodies were tested: the circular section cylinder and the Clark-Y aerodynamic profile. The Clark-Y aerodynamic profile and a circular section cylinder have very different flow configurations at the inlet. The Clark-Y profile is intended to increase lift and decrease drag, while the circular cylinder flow is more straightforward. Other differences between the two flow configurations are in the development of the boundary layer, pressure distribution, and flow separation patterns. For the Reynolds numbers tested, pressure profiles around the bodies were obtained for different angles α for the cylinder and β for the Clark-Y profile.

Circular section cylinder: As the cylinder has only one pressure point, pressures were taken varying the angle α and thus a distribution of pressure coefficients was generated, as a function of α, for each Reynolds number tested. To generate pressure coefficient distributions, varying the angle of attack (α) is essential since it mimics varied aerodynamic conditions, which aid in understanding the profile's performance under a range of Reynolds numbers and flow orientations. The inlet velocity (U1) was calculated using Equation 2.40 and the Reynolds was calculated using Equation 2.41, with the cylinder diameter being equal to the characteristic length. The inlet velocity (U1) is calculated by dividing the volumetric flow rate (Q) by the cross-sectional area (A) of the inlet. The Reynolds number (Re) is then determined using the formula $Re = (\rho * U1 * L) / \mu$, where ρ is the fluid density, U1 is the inlet velocity, L is the characteristic length, and μ is the dynamic viscosity. This method ensures that the flow conditions are correctly scaled for each test case, enabling accurate comparison and analysis. These data are shown in Table 3.

Table 3. Frequencies, speeds, and Reynolds numbers of cylinder tests

Frequency [Hz]	Pressure P0 [Pa]	Pressure P∞ [Pa]	Velocity [m/s]	Reynolds
7	-2.97	-23.77	5.89	14969
7.8	-2.97	-36.64	7.43	19418
10	-5.94	-59.43	9.45	24004
10.4	-6.93	-69.33	10.21	25928
13	-7.92	-128.75	14.14	36430

Thus, with the initial conditions and measured pressures in hand, the collected data were processed and the pressure coefficients for all measurement points were calculated. The figures (Figure 12 to Figure 14) show the pressure coefficient graphs obtained for the five Reynolds numbers tested.

Figure 12. Cylinder – Distribution of $\mathbf{C}_p$ *as a function of* α *for Reynolds equal to (A)* 1.5×10^4 *& (B)* 1.9×10^4

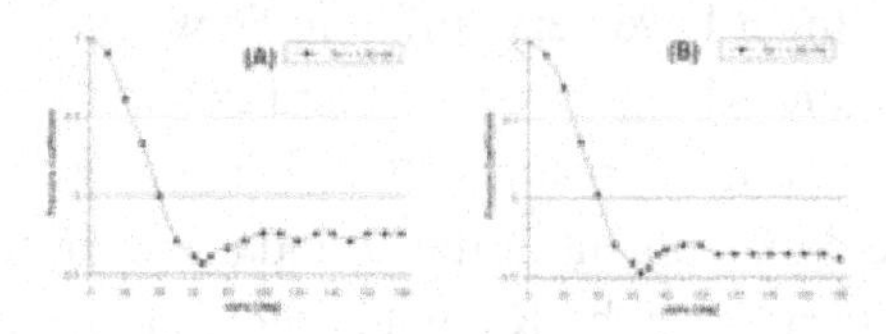

Figure 13. Cylinder – Distribution of $\mathbf{C}_p$ *as a function of* α *for Reynolds equal to (C)* 2.4×10^4 *& (D)*

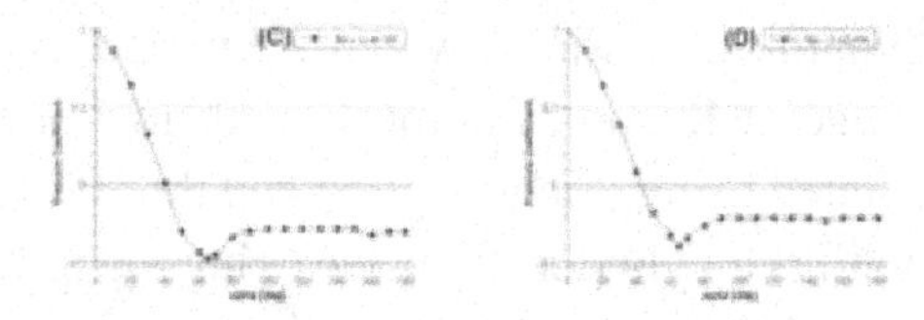

The results of the experiments resulted in pressure coefficient curves as a function of α in a format similar to the cases found in the literature. However, the values obtained were higher in modulus, which resulted in curves translated upwards, when compared to the curves found in the literature. Investigating this phenomenon raised the hypothesis that the surface roughness of the cylinder, previously considered negligible, was affecting the results. (Batham, J. P., et. al, 1973) carried out experiments with flow around a circular cylinder for both a smooth and a rough specimen. When comparing the results obtained for the cylinder in this work, with the results obtained by Batham, it was found that in fact, the results obtained in the wind tunnel were very similar to the curves related to the rough cylinder tests carried out by Batham. This finding corroborates the conclusion that, in fact, the discrepancies between the results obtained through the tests in this work and the curves provided by other researchers in the literature are mainly due to the fact that the test specimens used present non-negligible surface roughness.

Figure 14. Cylinder – Distribution of Cp as a function of α for Reynolds equal to

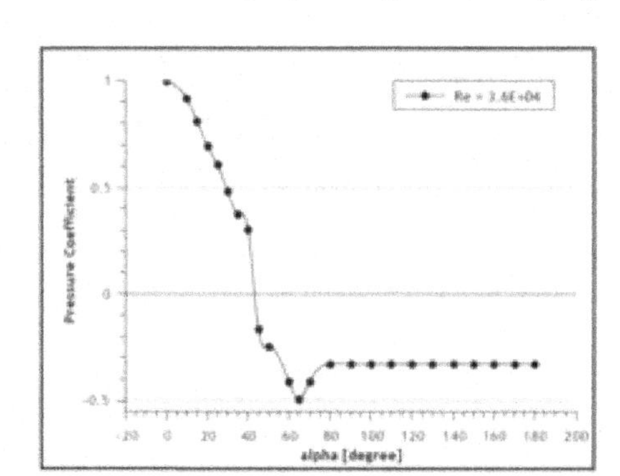

(2) Clark-Y aerodynamic profile

For each Reynolds number value and for each angle of attack tested, the pressures were taken at each of the seven points along the profile. With this, graphs were generated of the pressure coefficient around the Clark profile as a function of the ratio between the distance from the leading edge and the chord length x/L. In the experimental setup, the atmospheric pressure is measured using an Instrutemp brand barometer, which serves as a reference for calibrating other pressure measurement devices and guaranteeing data correctness. The input velocity (U1) was calculated using Equation 2.40 and the Reynolds was calculated, with the length of the profile chord being equal to the characteristic length.

These data are shown in Table 4.

Table 4. Frequencies, speeds, and Reynolds Numbers from the Clark Y tests

Frequency [Hz]	Pressure Po [Pa]	Pressure P∞ [Pa]	Velocity[m/s]	Reynolds
7	-7.84	-33.34	6.52	42108
8.5	-5.88	-41.19	7.67	49549
10	-9.81	-76.49	10.41	70099

The Clark-Y pressure coefficients were calculated for the tested conditions and their results are shown in the figure 15 to figure 24.

Figure 15. Clark-Y – as a Function of for Reynolds = and

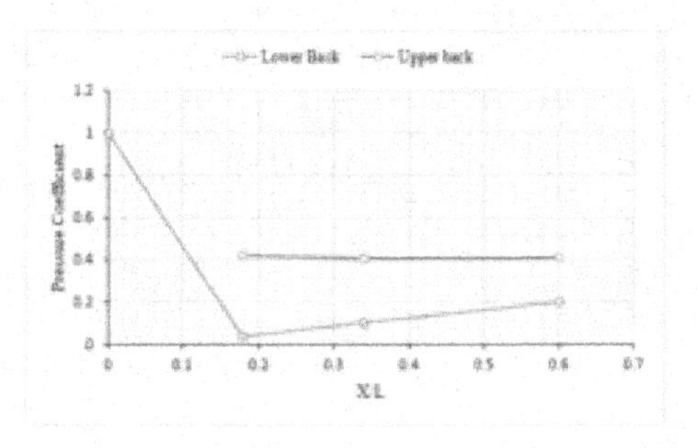

Figure 16. Clark-Y – as a Function of for Reynolds = and

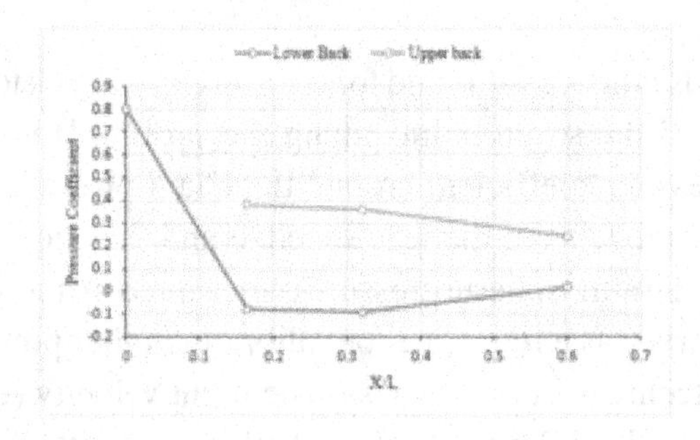

Figure 17. Clark-Y - as a Function of for Reynolds = and

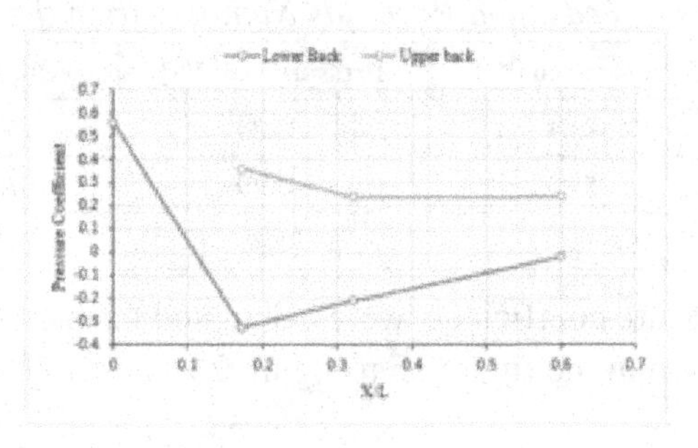

Figure 18. Clark-Y – as a Function of for Reynolds = and

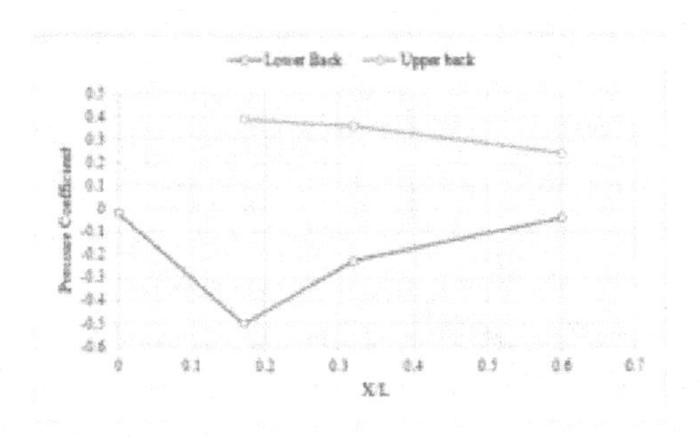

Figure 19. Clark-Y - as a Function of for Reynolds = and

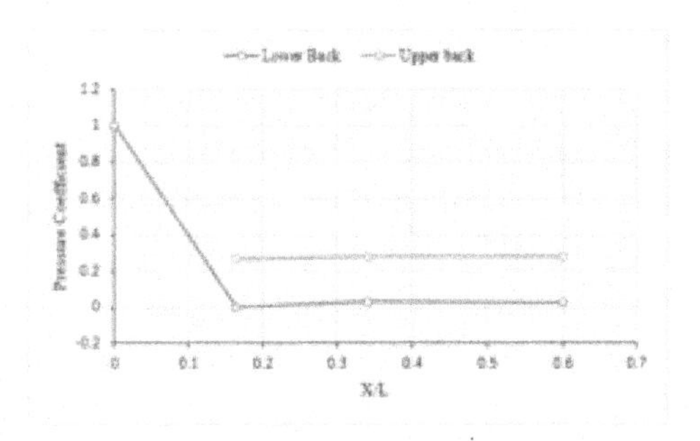

Figure 20. Clark-Y – as a Function of for Reynolds = and

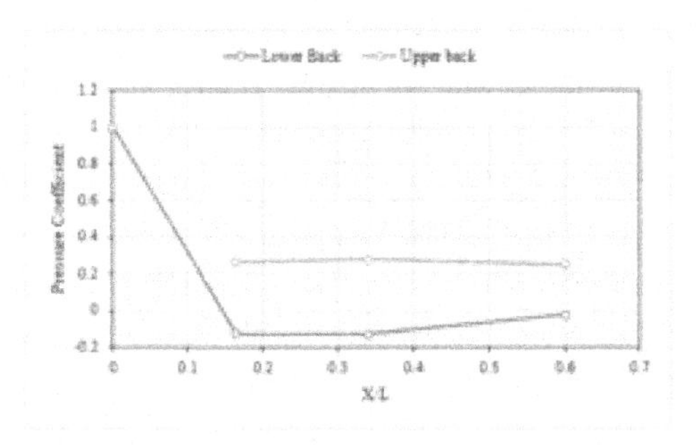

Figure 21. Clark-Y – as a Function of for Reynolds = and

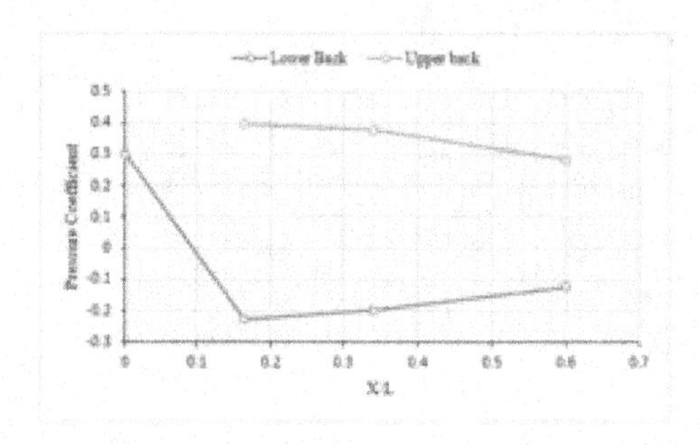

Figure 22. Clark-Y – as a Function of for Reynolds = and

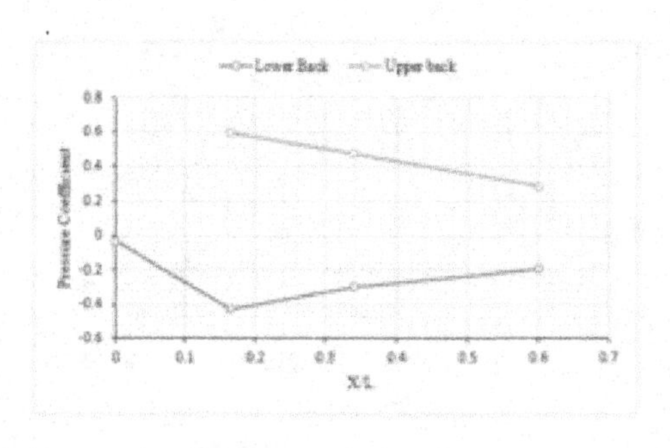

Figure 23. Clark-Y – as a Function of for Reynolds = and

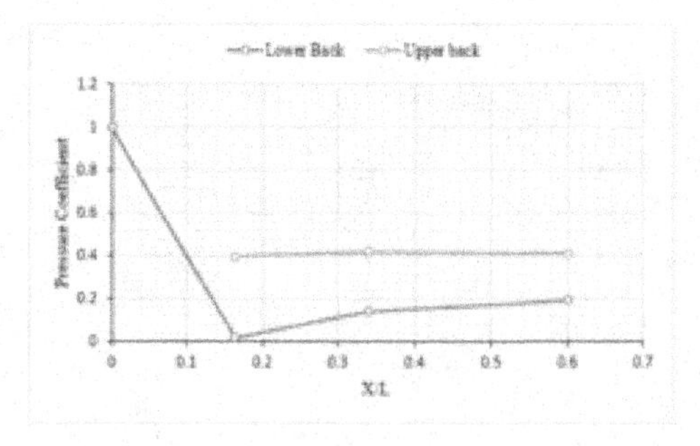

Figure 24. Clark-Y – as a Function of for Reynolds = and

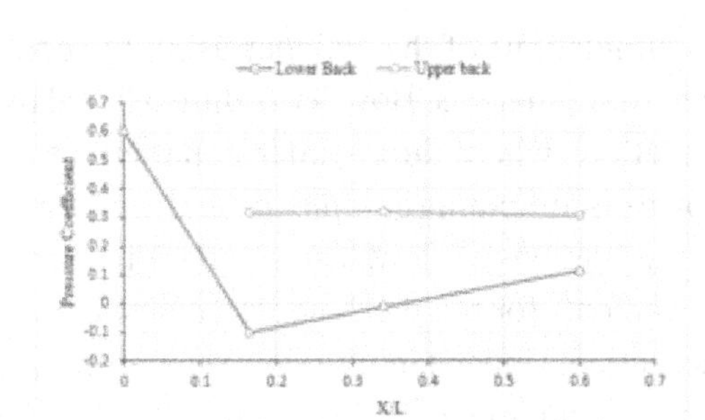

The graphs referring to the Clark-Y demonstrate the partial profile constructed from the seven points that the profile has for pressure taking. It can be noted that, for the same flow speed, the increase in the angle of attack causes the pressure coefficients on the upper back of the Clark-Y to decrease, while those on the lower back remain practically constant. Insights into lift generation, flow separation points, and stall characteristics were discovered through observations of variations in pressure coefficients along the upper and lower surfaces of the Clark-Y profile at various angles of attack. This data was crucial for validating and enhancing CFD models and aerodynamic designs. The increasing increase in the angle of attack, upon reaching a critical value, causes the phenomenon called stall, which consists of the loss of support on the profile. Increasing the angle of attack on a Clark-Y airfoil at constant flow speed can reduce pressure coefficients on its upper surface by promoting early flow separation, redistributing pressure gradients for less lift. However, excessively high angles can induce turbulent flow, increasing drag. Optimal angle of attack balancing lift and drag is essential for specific flight conditions.

It is also observed that for the same angle of attack, the pressure profile in the profile behaves practically indifferently to the increasing increase in flow velocity, i.e., in the Reynolds number. This occurred mainly because the Reynolds range in which the profile was tested, limited by the capacity of the instruments used, was very narrow, making it impossible to capture the real influences of this flow characteristic on the tested specimen. The study found that precise evaluation of flow characteristics' impact on specimen behavior was hindered by instrument constraints. These characteristics, like velocity and pressure distribution, are vital for understanding specimen performance.

RESULTS OF NUMERICAL SIMULATIONS

For the numerical simulations of the cylinder using the code, two-dimensional meshes were generated to simulate the flow. Two-dimensional meshes are useful for early research and situations where three-dimensional effects are negligible since they drastically cut down on computing complexity and time while retaining key flow features in numerical simulations of flow. The simulation returned, for the last time step, the pressure coefficients for the mesh nodes corresponding to the cylindrical profile with a circular section. Understanding flow separation, vortex formation, and drag characteristics require an understanding of the aerodynamic behavior, which may be obtained by generating the distribution of pressure coefficients within the circular section cylinder and identifying regions of high and low pressure.

The cylinder was simulated according to the conditions present in Table 2 and Table 3 for Reynolds numbers 3.6, using meshes with 440 and 1146 elements.

Figure 25. Computational time for simulations of the flow around the cylinder

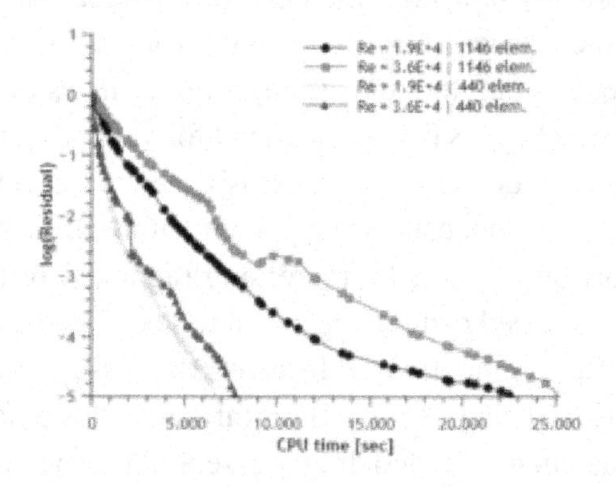

Figure 25 shows the results regarding the convergence time for each of the four simulated cases. For cases with the coarsest mesh, with 440 elements, the simulation time was 1.8 hours and 2.2 hours. For the cases in which the mesh with 1146 elements was used, the convergence time was much longer, being 6.4 hours and 7.0 hours. These results show that increasing the mesh refinement directly impacts the computational time required for the solution to converge, and that this relationship is not linear, as for a 2.6-fold increase in the number of mesh elements, the computational time increased by 3 to 4 times, approximately. Finer meshes take more elements and iterations to solve, which increases computational costs and lengthens simulation durations, but produce more accurate and detailed results. As a result, increasing mesh refinement directly affects computing time, frequently significantly.

CODE ANALYSIS AND VALIDATION

The simulations were performed for the two Reynolds numbers mentioned above, item 3.2. Two meshes were used: a mesh with 440 elements, the results of which are shown in figures (Figure 26 and 27), and a more refined mesh, with 1146 elements, with the results obtained present in figures (Figure 28 and Figure 29).

Figure 26. Cylinder – as a function of for Reynolds = 1 with the mesh of 440 elements

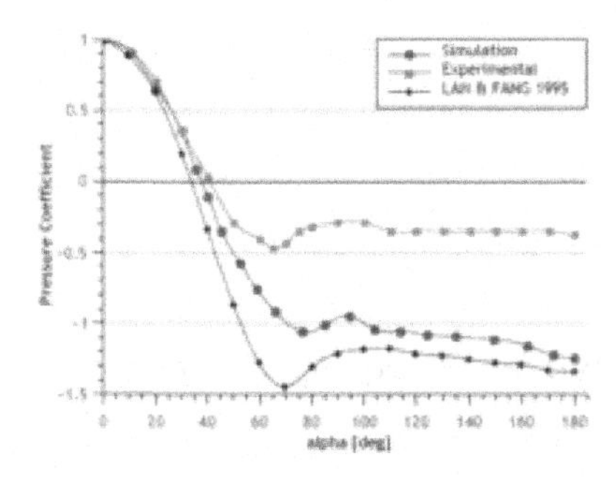

Figure 27. Cylinder – as a function of for Reynolds = with the mesh of 440 elements

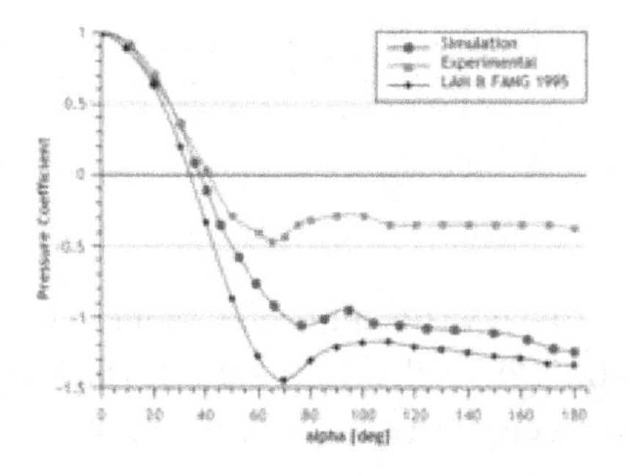

Figure 28. Cylinder – as a function of for Reynolds = 1 with the mesh of 1146 elements

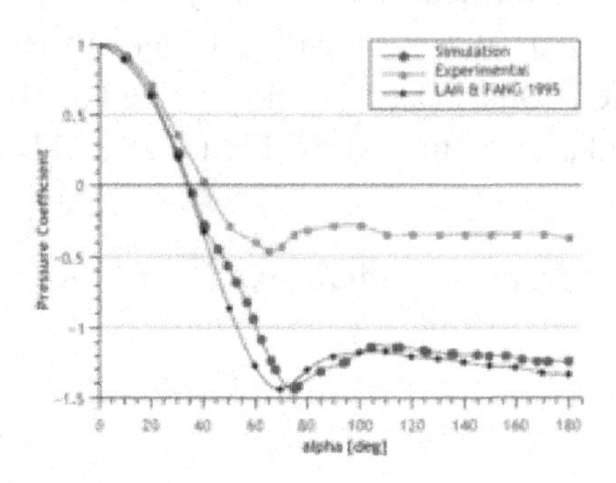

Figure 29. Cylinder – as a function of for Reynolds = 3.6 with the mesh of 1146 elements

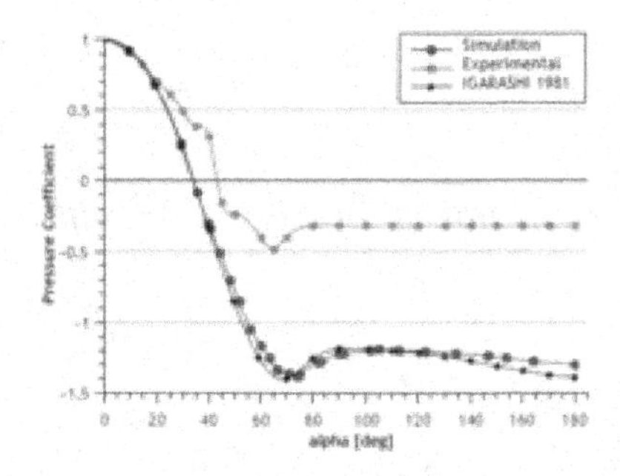

It is expected that at 0 the coefficient is one, as it coincides with the stagnation point, a point at which the flow velocity is zero. It is also expected that the minimum point of the curve occurs between 60 and 80, as it is in this sector that the detachment of the boundary layer occurs. After the minimum point, the coefficients are expected to increase and then stabilize at a value for angles greater than 100. Because of the advent of flow separation and turbulence, which produces considerable variations in pressure distributions and makes maintaining a steady value difficult, stabilizing the Cp value for angles more than 100° is often implausible, especially after the lowest point. The graphs show that in fact the shapes of the Cp curves as a function of α are physically coherent, as they all behave in a way that meets these theoretical-experimental expectations. Pressure coefficient (Cp) values play an important role in computational fluid dynamics (CFD) simulations because they offer a dimensionless measure of pressure variation along surfaces. This measure facilitates the identification of critical flow features, the evaluation of aerodynamic performance, and the validation of computational models against experimental or theoretical data.

The curves referring to the experimental tests did not coincide with those from the simulations and those from the literature due to the non-negligible roughness of the test specimens, however, it can be seen that the shape of the curves is similar to the others.

The graphs referring to the simulations using the more refined mesh show that the Cp values were closer to the literature values. Finer meshes better capture flow features and reduce numerical errors, hence improving the reliability of the simulation results. This is seen by the Cp values converge towards literature values when using a more refined mesh, strengthening the accuracy of the simulations. This shows that the code has good numerical accuracy, i.e., by refining the mesh the code improves the accuracy of the results.

CONCLUSION

The objectives of this work were to carry out experimental tests in a wind tunnel in order to obtain data for the validation of a computational code based on discretization in finite elements, use the code to simulate the cylindrical profile under the same test conditions and present a validation for the code.

The experimental tests were carried out for seventeen different conditions, five for the cylinder and twelve for the Clark-Y. The initial conditions were restricted to a Reynolds range around due to instrumental limitations, in this case the inclined column manometer, but still below a critical Reynolds for which the flow would transition to the turbulent regime.

The pressure coefficient curves obtained from the cylinder tests in the laboratory had a format similar to the cases in the literature: CP equal to one at stagnation and a minimum point at approximately 70, which is where the detachment of the boundary layer occurs. Higher mesh resolution improves result accuracy by collecting finer features of the flow, lowering numerical errors, and producing more dependable pressure distribution data, as evidenced by the improvement in Cp values with mesh refinement. However, the module values of the coefficients became larger. It was verified, through analysis of a similar work, Batham (1973), that the physical occurrence shown in the results of the present work is mainly related to the non-negligible surface roughness of the specimen.

The Clark-Y tests resulted in partial pressure profiles around the body. Extrapolating the trend lines of the graphs, it was noticed that the curves had a shape, in general, similar to curves of other Clark profiles tested in the literature. The lower back has higher pressure coefficients than the upper, which in fact occurs, as the speed on the upper back is greater, and this difference in pressure is what results in the support force. In order to create pressure coefficient graphs, pressures are measured throughout the profile, normalized by the dynamic pressure, and then Cp is plotted as a function of both the chord length ratio (x/L) and the distance from the leading edge.

Numerical simulations were carried out for the cylindrical specimen. The two conditions for which data were found in the literature were simulated and experimental tests were carried out. From the simulation results, pressure coefficient graphs around the cylinder were obtained. The curves resulting from the simulations did not coincide with those obtained through experimental tests, for the reason previously explained. In order to ensure consistency and comparability of results, certain parameters are maintained constant across simulations and experimental tests. These parameters include fluid characteristics (density and viscosity), boundary conditions, inlet velocity, and geometric configurations. However, the pressure coefficient curves resulting from the simulations fit reasonably well with those originating from literature data.

In order to assist future work, in addition to the experimental results referring to the Clark-Y specimen, discretization's of the profile were also prepared from the specimen for the four angles of attack tested, as the Clark-Y does not have an algebraic equation that can generate its curvature, thus facilitating the creation of meshes so that this body can also be simulated in the future.

REFERENCES

Batham, J. P. (1973). Pressure distributions on circular cylinders at critical Reynolds numbers. *Journal of Fluid Mechanics*, 57(2), 209–228. 10.1017/S0022112073001114

Bazant, M. Z. (2016). Exact solutions and physical analogies for unidirectional flows. *Physical Review Fluids*, 1(2), 024001. 10.1103/PhysRevFluids.1.024001

Bhalekar, S., & Patade, J. (2016). Analytical solutions of nonlinear equations with proportional delays. *Applied and Computational Mathematics*, 15(3), 331–345.

Chattouh, A. (2022). Numerical solution for a class of parabolic integro-differential equations subject to integral boundary conditions. *Arabian Journal of Mathematics*, 11(2), 213–225. 10.1007/s40065-022-00371-3

Fernandes, J. W. D. (2014). *Development of a program for simulating incompressible flows with a moving boundary. 62 p. Monograph (Course completion work).* Federal Technological University of Paraná, Pato Branco.

Fox, R. W., Mcdonald, A. T., & Pritchard, P. J. (2014). *Introduction to fluid mechanics*. 8th ed.) Gen-LTC.

Geuzaine, C., & Remacle, J. F. (2009). Gmsh: A 3-D finite element mesh generator with built-in pre-and post-processing facilities. *International Journal for Numerical Methods in Engineering*, 79(11), 1309–1331. 10.1002/nme.2579

Heidari, E., Daeichian, A., Movahedirad, S., & Sobati, M. A. (2022). Analysis of collision characteristics in a 3D gas-solid tapered fluidized bed. *Chemical Engineering Research & Design*, 184, 554–565. 10.1016/j.cherd.2022.06.002

Igarashi, T. (1981). Characteristics of the flow around two circular cylinders arranged in tandem: 1st report. *Bulletin of the JSME*, 24(188), 323–331. 10.1299/jsme1958.24.323

Lam, K., & Fang, X. (1995). The effect of interference of four equispaced cylinders in cross flow on pressure and force coefficients. *Journal of Fluids and Structures*, 9(2), 195–214. 10.1006/jfls.1995.1010

Reddy, J. N. (2005). *An introduction to the finite element method* (Vol. 3). McGraw-Hill.

Tamrakar, A., Devarampally, D. R., & Ramachandran, R. (2018). Advanced multiphase hybrid model development of fluidized bed wet granulation processes. [). Elsevier.]. *Computer-Aided Chemical Engineering*, 41, 159–187. 10.1016/B978-0-444-63963-9.00007-5

Wang, A. X., & Kabala, Z. J. (2022). Body Morphology and Drag in Swimming: CFD Analysis of the Effects of Differences in Male and Female Body Types. *Fluids (Basel, Switzerland)*, 7(10), 332. 10.3390/fluids7100332

Compilation of References

Al-Hamido, R. K. (2019). *A study of multi-Topological Spaces* [PhD Thesis, AlBaath University].

Al-Hamido, R. K. (2023). Infra Bi-topological Spaces. *Prospects for Applied Mathematics and Data Analysis, 1*, 8-16.

Al-Hamido, R. K. (2018). Neutrosophic Crisp Bi-Topological Spaces. *Neutrosophic Sets and Systems*, 21, 66–73.

Al-Odhari. (2015). On infra topological space. *International Journal of Mathematical Archive*, 11(5), 179–184.

Al-Ratrout, S., Tarawneh, O. H., Altarawneh, M. H., & Altarawneh, M. Y. (2019). Mobile application development methodologies adopted in Omani Market: A comparative study. *International Journal of Software Engineering and Its Applications*, 10(2).

Al-Shami, T. M. (2017a). On supra semi open sets and some applications on topological spaces. *Journal of Advanced Studies in Topology*, 8(2), 144–153. 10.20454/jast.2017.1335

Al-Shami, T. M. (2017b). Utilizing supra α-open sets to generate new types of supra compact and supra lindel of spaces, Facta Universitatis, Series. *Mathematics and Informatics*, 32, 151–162.

Al-Shami, T. M. (2018). Supra semi-compactness via supra topological spaces. *Journal of Taibah University for Science*, 12(3), 338–343. 10.1080/16583655.2018.1469835

Al-Shami, T. M., Asaad, B. A., & El-Gayar, M. A. (2020). *Various types of supra pre-compact and supra pre-Lindel¨of spaces* (Vol. 32). Missouri Journal of Mathematical Science.

Argyris, J. H., & Kelsey, S. (1960). *Energy theorems and structural analysis* (Vol. 60). Butterworths. 10.1007/978-1-4899-5850-1

Arnold, M., Burgermeister, B., Führer, C., Hippmann, G., & Rill, G. (2011). Numerical methods in vehicle system dynamics: State of the art and current developments. *Vehicle System Dynamics*, 49(7), 1159–1207. 10.1080/00423114.2011.582953

Asiedu, Chapman-Wardy, & Doku-Amponsah. (2021). *A Modification of Newton Method for Solving Non-linear Equations.* Academic Press.

Ayada, W. M., & Hammad, M. A. E. E. (2023). Design quality criteria for smartphone applications interface and its impact on user experience and usability. *International Design Journal*, 13(4), 339–354. 10.21608/idj.2023.305364

Azure, I., Aloliga, G., & Doabil, L. (2019). Comparative Study of Numerical Methods for Solving Non-linear Equations Using Manual Computation. *Mathematics Letters.*, 5(4), 41. 10.11648/j.ml.20190504.11

Bahrudin, I. A., Mohd Hanifa, R., Abdullah, M. E., & Kamarudin, M. F. (2013). Adapting extreme programming approach in developing electronic document online system (eDoc). *Applied Mechanics and Materials*, 321, 2938–2941. 10.4028/www.scientific.net/AMM.321-324.2938

Barton, N. G. (2013). Simulations of air-blown thermal storage in a rock bed. *Applied Thermal Engineering*, 55(1-2), 43–50. 10.1016/j.applthermaleng.2013.03.002

Batham, J. P. (1973). Pressure distributions on circular cylinders at critical Reynolds numbers. *Journal of Fluid Mechanics*, 57(2), 209–228. 10.1017/S0022112073001114

Bazant, M. Z. (2016). Exact solutions and physical analogies for unidirectional flows. *Physical Review Fluids*, 1(2), 024001. 10.1103/PhysRevFluids.1.024001

Bhalekar, S., & Patade, J. (2016). Analytical solutions of nonlinear equations with proportional delays. *Applied and Computational Mathematics*, 15(3), 331–345.

Bidwell, J. K. (1994). Archimedes and Pi-Revisited. *School Science and Mathematics*, 94(3), 127–129. 10.1111/j.1949-8594.1994.tb15638.x

Booker, K., & Nec, Y. (2019). On accuracy of numerical solution to boundary value problems on infinite domains with slow decay. *Mathematical Modelling of Natural Phenomena*, 14(5), 503. 10.1051/mmnp/2019008

Borras, H., Duran, R., & Iriarte, R. (1984). *Notes on numerical methods*. UNAM Engineering Faculty.

Boutayeb, A., & Abdelaziz, C. (2007). A mini-review of numerical methods for high-order problems. *International Journal of Computer Mathematics*, 84(4), 563–579. 10.1080/00207160701242250

Boyce, W. E. & DiPrima, R. C. (1970). *Elemental Differential Equations and Boundary Value Problems*. J. Wiley.

Cacciola, P., & Tombari, A. (2021). Steady state harmonic response of nonlinear soil-structure interaction problems through the Preisach formalism. *Soil Dynamics and Earthquake Engineering*, 144, 106669. 10.1016/j.soildyn.2021.106669

Carson, H. A., Darmofal, D. L., Galbraith, M. C., & Allmaras, S. R. (2017). Analysis of output-based error estimation for finite element methods. *Applied Numerical Mathematics*, 118, 182–202. 10.1016/j.apnum.2017.03.004

Cascetta, M., Cau, G., Puddu, P., & Serra, F. (2016). A comparison between CFD simulation and experimental investigation of a packed-bed thermal energy storage system. *Applied Thermal Engineering*, 98, 1263–1272. 10.1016/j.applthermaleng.2016.01.019

Chattouh, A. (2022). Numerical solution for a class of parabolic integro-differential equations subject to integral boundary conditions. *Arabian Journal of Mathematics*, 11(2), 213–225. 10.1007/s40065-022-00371-3

Chegeni, Sharbatdar, & Mahjoub, &Raftari. (2022). *Numerical Methods in Civil Engineering A novel method for detecting structural damage based on data-driven and similarity-based techniques under environmental and operational changes*. Numerical Methods in Civil Engineering. 10.52547/nmce.6.4.16

Chomal, V. S., & Saini, J. R. (2014). Significance of software documentation in software development process. *International Journal of Engineering Innovations and Research*, 3(4), 410.

Clough, R. W. (1960). *The finite element in plane stress analysis. Proc. 2^< nd> ASCE Confer.* On Electric Computation.

Codina, R., Morton, C., Oñate, E., & Soto, O. (2000). Numerical aerodynamic analysis of large buildings using a finite element model with application to a telescope building. *International Journal of Numerical Methods for Heat & Fluid Flow*, 10(6), 616–633. 10.1108/09615530010347196

Constantinescu, R., & Iacob, I. M. (2007). Capability maturity model integration. *Journal of Applied Quantitative Methods*, 2(1), 31–37.

Coorevits, P., & Bellenger, E. (2004). Alternative mesh optimality criteria for h-adaptive finite element method. *Finite Elements in Analysis and Design*, 40(9-10), 1195–1215. 10.1016/j.finel.2003.08.007

Darmawan, R. N. (2016). Comparison of the Gauss-Legendre, Gauss-Lobatto, and Gauss-Kronrod Methods for Numerical Integration of Exponential Functions. *Journal of Mathematics and Mathematics Education*, 1(2), 99–108.

De Carvalho, B. V., & Mello, C. H. P. (2011). Scrum agile product development method-literature review, analysis and classification. *Product: Management & Development*, 9(1), 39–49. 10.4322/pmd.2011.005

Dodgson, M., Gann, D. M., & Salter, A. (2007). The impact of modelling and simulation technology on engineering problem solving. *Technology Analysis and Strategic Management*, 19(4), 471–489. 10.1080/09537320701403425

Drmač, Z. (2020). Numerical methods for accurate computation of the eigenvalues of Hermitian matrices and the singular values of general matrices. *SeMA Journal.*, 78(1), 53–92. Advance online publication. 10.1007/s40324-020-00229-8

Dybå, T., Dingsøyr, T., & Moe, N. B. (2014). Agile project management. *Software project management in a changing world*, 277-300.

Erhunmwun, I. D., & Ikponmwosa, U. B. (2017). Review on finite element method. *Journal of Applied Science & Environmental Management*, 21(5), 999–1002. 10.4314/jasem.v21i5.30

Ermawati, E., Rahayu, P., & Zuhairoh, F. (2017). Comparison of Numerical Solutions for Double Integrals in Algebraic Functions using the Romberg Method and Monte Carlo Simulation. *MSA Journal*, 5(1), 46–57.

Eymard, R., Thierry, G., & Herbin, R. (2000). Finite volume methods. *Finite Volume Methods.*, 7, 713–1018. 10.1016/S1570-8659(00)07005-8

Ezzatabadipour, M., & Zahedi, H. (2018). Simulation of a fluid flow and investigation of a permeability-porosity relationship in porous media with random circular obstacles using the curved boundary lattice Boltzmann method. *The European Physical Journal Plus*, 133(11), 464. 10.1140/epjp/i2018-12325-2

Fadrique Ruano, G. (2017). *Design of a ventilation system for fan testing*. Academic Press.

Fei, Y.-F., Tian, Y., Huang, Y., & Lu, X. (2022). Influence of damping models on dynamic analyses of a base-isolated composite structure under earthquakes and environmental vibrations. Gong Cheng Li Xue. *Gongcheng Lixue*, 39, 201–211. 10.6052/j.issn.1000-4750.2021.07.0500

Fernandes, J. W. D. (2014). *Development of a program for simulating incompressible flows with a moving boundary. 62 p. Monograph (Course completion work).* Federal Technological University of Paraná, Pato Branco.

Finkelshtein, A. M. (n.d.). The numerical analysis in the last 24 years. *Magazine of the Spanish Society of History of Sciences and Techniques*, 26, 919–928.

Finlayson, B. A., & Scriven, L. E. (1965). The method of weighted residuals and its relation to certain variational principles for the analysis of transport processes. *Chemical Engineering Science*, 20(5), 395–404. 10.1016/0009-2509(65)80052-5

Fox, R. W., Mcdonald, A. T., & Pritchard, P. J. (2014). *Introduction to fluid mechanics*. 8th ed.) Gen-LTC.

Ge, G., Wang, X., Manzano, J., & Gao, G. (2009). *Tile Percolation: An OpenMP Tile Aware Parallelization Technique for the Cyclops-64 Multicore Processor*. Springer. .10.1007/978-3-642-03869-3_78

Geuzaine, C., & Remacle, J. F. (2009). Gmsh: A 3-D finite element mesh generator with built-in pre-and post-processing facilities. *International Journal for Numerical Methods in Engineering*, 79(11), 1309–1331. 10.1002/nme.2579

Graciano-Uribe, J., Pujol, T., Puig-Bargués, J., Duran-Ros, M., Arbat, G., & Ramírez de Cartagena, F. (2021). Assessment of Different Pressure Drop-Flow Rate Equations in a Pressurized Porous Media Filter for Irrigation Systems. *Water (Basel)*, 13(16), 2179. 10.3390/w13162179

Guicciardini, N., Kjeldsen, T. H., & Rowe, D. E. (2006). Mathematics in the Physical Sciences, 1650-2000. *Oberwolfach Reports*, 2(4), 3175–3246. 10.4171/owr/2005/56

Guimarães, F., Saldanha, R., Mesquita, R., Lowther, D., & Ramirez, J. (2007). A Meshless Method for Electromagnetic Field Computation Based on the Multiquadric Technique. Magnetics. *IEEE Transactions on Magnetics*, 43(4), 1281–1284. 10.1109/TMAG.2007.892396

Heidari, E., Daeichian, A., Movahedirad, S., & Sobati, M. A. (2022). Analysis of collision characteristics in a 3D gas-solid tapered fluidized bed. *Chemical Engineering Research & Design*, 184, 554–565. 10.1016/j.cherd.2022.06.002

Hema, V., Thota, S., Kumar, S. N., Padmaja, C., Krishna, C. B. R., & Mahender, K. (2020, December). Scrum: An effective software development agile tool. *IOP Conference Series. Materials Science and Engineering*, 981(2), 022060. 10.1088/1757-899X/981/2/022060

Herfina, N., Amrullah, A., & Junaidi, J. (2019). The Effectiveness of the Trapezoidal and Simpson Methods in Determining Area Using Pascal Programming. *Mandalika Mathematics and Education Journal*, 1(1), 53–65. 10.29303/jm.v1i1.1242

Huang, J., Yang, R., Ge, H., & Tan, J. (2021). An effective determination of the minimum circumscribed circle and maximum inscribed circle using the subzone division approach. *Measurement Science & Technology*, 32(7), 075014. 10.1088/1361-6501/abf803

Idelsohn, S., Oñate, E., Calvo, N., & Del Pin, F. (2003). Meshless finite element method. *International Journal for Numerical Methods in Engineering, 58*, 893-912. .10.1002/nme.798

Igarashi, T. (1981). Characteristics of the flow around two circular cylinders arranged in tandem: 1st report. *Bulletin of the JSME*, 24(188), 323–331. 10.1299/jsme1958.24.323

Imanova, G. (2022). History of Physics. *Journal of Physics & Optics Sciences.* doi. org/ 10.47363/ JPSOS/2022

Isaacson, W. (2014). The innovators: How a group of hackers, geniuses, and geeks created the digital revolution. *Journal of Multidisciplinary Research*, 7(1), 111.

Jagota, V., Sethi, A., & Kumar, D.-K. (2013). Finite Element Method: An Overview. *Walailak Journal of Science and Technology*, 10, 1–8. 10.2004/wjst.v10i1.499

Janna, W. S. (2020). *Introduction to fluid mechanics.* CRC press.

Javaid, M., Haleem, A., Singh, R. P., Suman, R., & Gonzalez, E. S. (2022). Understanding the adoption of Industry 4.0 technologies in improving environmental sustainability. *Sustainable Operations and Computers*, 3, 203–217. 10.1016/j.susoc.2022.01.008

Jayaparthasarathy, G., Little Flower, V. F., & Arockia Dasan, M. (2023). Neutrosophic Supra Topological Applications in Data Mining Process. *Neutrosophic Sets and Systems, 27*, 80-97.

Kahourzade, S., Mahmoudi, A., Gandomkar, A., Rahim, A., Ping, H. W., & Uddin, M. N. (2013). Design optimization and analysis of AFPM synchronous machine incorporating power density, thermal analysis, and dback-EMF THD. *Electromagnetic Waves*, 136, 327–367. 10.2528/ PIER12120204

Kalies, G., & Do, D. D. (2023). Momentum work and the energetic foundations of physics. IV. The essence of heat, entropy, enthalpy, and Gibbs free energy. *AIP Advances*, 13(9), 095126. 10.1063/5.0166916

Kasper, E. (2001). The Finite-Difference Method (FDM). *Advances in Imaging and Electron Physics*, 116, 115–191. 10.1016/S1076-5670(01)80068-9

Kasper, E. (2018). The Finite-Difference Method (FDM). In *Advances in Imaging and Electron Physics*. Elsevier.

Katsikadelis, J. (2002). *Boundary Elements. Theory and Applications*. Elsevier.

Keyser, M. J., Conradie, M., Coertzen, M., & Vandyk, J. (2006). Effect of Coal Particle Size Distribution on Packed Bed Pressure Drop and Gas Flow Distribution. *Fuel*, 85(10-11), 1439–1445. 10.1016/j.fuel.2005.12.012

Kim, A., Park, C., & Park, S.-H. (2003). Development of Web-based Engineering Numerical Software (WENS) Using MATLAB: Applications to linear algebra. *Computer Applications in Engineering Education*, 11(2), 67–74. 10.1002/cae.10038

Kythe, P., Wei, D., & Okrouhlik, M. (2004). An Introduction to Linear and Nonlinear Finite Element Analysis: A Computational Approach. Applied Mechanics Reviews -. *Applied Mechanics Reviews*, 57(5), B25. Advance online publication. 10.1115/1.1818688

Lalsing, V., Kishnah, S., & Pudaruth, S. (2012). People factors in agile software development and project management. *International Journal of Software Engineering and Its Applications*, 3(1), 117–137. 10.5121/ijsea.2012.3109

Lam, K., & Fang, X. (1995). The effect of interference of four equispaced cylinders in cross flow on pressure and force coefficients. *Journal of Fluids and Structures*, 9(2), 195–214. 10.1006/jfls.1995.1010

Langlois & Deville. (2014). *Introduction to the Finite Element Method.* Springer. .10.1007/978-3-319-03835-3_10

Li, G. (2005). Meshless Methods for Numerical Solution of Partial Differential Equations. 10.1007/978-1-4020-3286-8_128

Li, G., & Yu, D.-H. (2018). Efficient Inelasticity-Separated Finite-Element Method for Material Nonlinearity Analysis. *Journal of Engineering Mechanics*, 144(4), 04018008. 10.1061/(ASCE)EM.1943-7889.0001426

Li, J., Guan, Y., Wang, G., Wang, G., Zhang, H., & Lin, J. (2020, June). A meshless method for topology optimization of structures under multiple load cases. In *Structures* (Vol. 25, pp. 173–179). Elsevier. 10.1016/j.istruc.2020.03.005

Li, K., Jarrar, F., Sheikh-Ahmad, J., & Ozturk, F. (2017). Using coupled Eulerian Lagrangian formulation for accurate modeling of the friction stir welding process. *Procedia Engineering*, 207, 574–579. 10.1016/j.proeng.2017.10.1023

Liu, Zhang, Su, Sun, Han, & Wang. (2020). Convergence Analysis of Newton-Raphson Method in Feasible Power-Flow for DC Network. *IEEE Transactions on Power Systems*. IEEE. .10.1109/TPWRS.2020.2986706

Long, H., Wang, Z., Zhang, C., Zhuang, H., Chen, W., & Peng, C. (2021). Nonlinear study on the structure-soil-structure interaction of seismic response among high-rise buildings. *Engineering Structures*, 242, 112550. 10.1016/j.engstruct.2021.112550

Mahmudova, S. (2018). *Features of Programming Languages and Algorithm for Calculating the Effectiveness*. Academic Press.

Mahomed, N., & Kekana, M. (1998). An error estimator for adaptive mesh refinement analysis based on strain energy equalisation. *Computational Mechanics*, 22(4), 355–366. 10.1007/s004660050367

Mashhour, S., Allam, A. A., Mahmoud, F. S., & Khedr, F. H. (1983). On Supra topological spaces. *Indian Journal of Pure and Applied Mathematics*, 14(4), 502–510.

McCracken, D., & Dorn, W. (1984). *Numerical methods and Fortran programming* (Limusa, Ed.). Academic Press.

Merci, B. (2016). Introduction to fluid mechanics. *SFPE Handbook of Fire Protection Engineering*, 1–24. SFPE.

Mhemdia, T. M. (2024). Al-shami, Introduction to temporal topology. *J. Math. Computer Sci.*, 34(3), 205–217. 10.22436/jmcs.034.03.01

Milici, C., Draganescu, G., & Machado, T. (2019). Numerical Methods. 10.1007/978-3-030-00895-6_

Murakami, A., Wakai, A., & Fujisawa, K. (2010). Numerical Methods. *Soil and Foundation*, 50(6), 877–892. 10.3208/sandf.50.877

Nagarajan, P. (2018). *Matrix Methods of Structural Analysis* (1st ed.). CRC Press. 10.1201/9781351210324

Nasution, M. K. (2020). *Industry 4.0*. Academic Press.

Navarro, C., Hitschfeld, N., & Mateu, L. (2013). A Survey on Parallel Computing and its Applications in Data-Parallel Problems Using GPU Architectures. *Communications in Computational Physics*, 15(2), 285–329. 10.4208/cicp.110113.010813a

Nemec, D., & Levec, J. (2005). Flow Through Packed Bed Reactors: 1. Single-Phase Flow. *Chemical Engineering Science*, 60(24), 6947–6957. 10.1016/j.ces.2005.05.068

Nguyen, V. P., Rabczuk, T., Bordas, S., & Duflot, M. (2008). Meshless methods: A review and computer implementation aspects. *Mathematics and Computers in Simulation*, 79(3), 763–813. 10.1016/j.matcom.2008.01.003

Nochetto, R. H., Siebert, K. G., & Veeser, A. (2009). Theory of adaptive finite element methods: an introduction. In *Multiscale, Nonlinear and Adaptive Approximation: Dedicated to Wolfgang Dahmen on the Occasion of his 60th Birthday* (pp. 409-542). Springer Berlin Heidelberg. 10.1007/978-3-642-03413-8_12

Oñate, E. (2013). *Structural analysis with the finite element method. Linear statics: volume 2: beams, plates and shells*. Springer Science & Business Media.

Oñate, E., & Bugeda, G. (1993). A study of mesh optimality criteria in adaptive finite element analysis. *Engineering Computations, 10*(4), 307-321. .10.1108/eb023910

Ozturk, T. Y., & Ozkan, A. (2019). Neutrosophic Bi-topological Spaces. *Neutrosophic Sets and Systems*, 30, 88–97.

Pacheco, J. P. F. R., da Silva, J., & do Vale, J. L. (2020). *h-adaptative strategy proposal for Finite Element Method applied in structures imposed to multiple load cases*. Academic Press.

Papazafeiropoulos, G. (2016). Newton Raphson Line Search - Program for the solution of equations with the quasi-Newton-Raphson method accelerated by a line search algorithm. 10.13140/RG.2.1.3485.5282

Patel, V. G., & Rachchh, N. V. (2020). Meshless method–review on recent developments. *Materials Today: Proceedings*, 26, 1598–1603. 10.1016/j.matpr.2020.02.328

Pepper, D. (2005). *The Finite Element Method: Basic Concepts and Applications*. Taylor & Francis. .10.1201/9780203942352

Perbani, N. M. R. R. C., & Rinaldy. (2018). Application of Simpson's Volume Calculation Method to Calculate Ship Volume and Land Topography. *Journal of Green Engineering*, 2(1), 90–100.

Pesic, R., & Radoičić, K. (2015). Pressure drop in packed beds of spherical particles at ambient and elevated air temperatures. *Chemical Industry & Chemical Engineering Quarterly*, 21(3), 419–427. 10.2298/CICEQ140618044P

Pomeroy-Huff, M., Mullaney, J. L., Cannon, R., & Seburn, M. (2005). *The Personal Software Process (PSP) Body of Knowledge, Version 1.0.*

Ponalagusamy, Pja, & Muthusamy. (2010). Numerical Methods on Ordinary Differential Equation. *International Conference on Emerging Trends in Mathematics and Computer Applications 2010, 1.*

Pradhan & Snehashish. (2019). Finite Element Method. In *Computational Structural Mechanics*. Academic Press.

Pranto, M. R. I., & Inam, M. I. (2020). Numerical Analysis of the Aerodynamic Characteristics of NACA4312 Airfoil. *Journal of Engineering Advancements*, 1(02), 29–36. 10.38032/jea.2020.02.001

Purnama, I. (2023). Clinical Information System Using Extreme Programming Method. International Journal of Science. *Technology & Management*, 4(5), 1229–1235.

Qi, E. S., Bi, H. T., & Bi, X. (2019). Study on agile software development based on scrum method. In *Proceeding of the 24th International Conference on Industrial Engineering and Engineering Management 2018* (pp. 430-438). Springer Singapore. 10.1007/978-981-13-3402-3_46

Rahman, M. M. (2021). Understanding Science and Preventing It from Becoming Pseudoscience. *Philosophy (London, England)*, 9(3), 127–135.

Reddy, J. N. (2005). *An introduction to the finite element method* (Vol. 3). McGraw-Hill.

Rionda, S. B. (n.d.). *Examples of application of numerical methods to engineering problems.* A.C. Mathematics Research Center.

Russell, A. (2016, January-March). The innovators: How a group of hackers, geniuses, and geeks created the digital revolution (Isaacson, W.; 2014) [book review]. *IEEE Annals of the History of Computing*, 38(1), 94–c3. 10.1109/MAHC.2016.8

Sandoval-Hernandez, M. A., Vazquez-Leal, H., Filobello-Nino, U., & Hernandez-Martinez, L. (2019). New handy and accurate approximation for the Gaussian integrals with applications to science and engineering. *Open Mathematics*, 17(1), 1774–1793. 10.1515/math-2019-0131

Schenk, O., & Gärtner, K. (2002). Two-level dynamic scheduling in PARDISO: Improved scalability on shared memory multiprocessing systems. *Parallel Computing*, 28(2), 187–197. 10.1016/S0167-8191(01)00135-1

Shafiee, S., Wautelet, Y., Hvam, L., Sandrin, E., & Forza, C. (2020). Scrum versus Rational Unified Process in facing the main challenges of product configuration systems development. *Journal of Systems and Software*, 170, 110732. 10.1016/j.jss.2020.110732

Shawkat, S. A., Ismail, R. N., & Abdulqader, I. R. (2022, November). Implementation a hybrid ADHOC sensor system. In *AIP Conference Proceedings* (*Vol. 2394*, No. 1). AIP Publishing. 10.1063/5.0121146

Shrivastava, A., Jaggi, I., Katoch, N., Gupta, D., & Gupta, S. (2021, July). A systematic review on extreme programming. *Journal of Physics: Conference Series*, 1969(1), 012046. 10.1088/1742-6596/1969/1/012046

Soomro, S., Shaikh, A., Qureshi, S., & Ali, B. (2023). A Modified Hybrid Method For Solving Non-Linear Equations With Computational Efficiency. *VFAST Transactions on Mathematics.*, 11(2), 126–137. 10.21015/vtm.v11i2.1620

Tamrakar, A., Devarampally, D. R., & Ramachandran, R. (2018). Advanced multiphase hybrid model development of fluidized bed wet granulation processes. [). Elsevier.]. *Computer-Aided Chemical Engineering*, 41, 159–187. 10.1016/B978-0-444-63963-9.00007-5

Tuama, B. A., Shawkat, S. A., & Askar, N. A. (2022, November). Recognition and classification of facial expressions using artificial neural networks. In *Proceedings of Third Doctoral Symposium on Computational Intelligence: DoSCI 2022* (pp. 229-246). Singapore: Springer Nature Singapore.

Vulandari, R. T. (2017). *Numerical Methods: Theory, Cases, and Applications.* Mavendra Pres.

Wang, A. X., & Kabala, Z. J. (2022). Body Morphology and Drag in Swimming: CFD Analysis of the Effects of Differences in Male and Female Body Types. *Fluids (Basel, Switzerland)*, 7(10), 332. 10.3390/fluids7100332

Winter, R., Valsamidou, A., Class, H., & Flemisch, B. (2022). A Study on Darcy versus Forchheimer Models for Flow through Heterogeneous Landfills Including Macropores. *Water (Basel)*, 14(4), 546. 10.3390/w14040546

Wu, N. J., Chen, B. S., & Tsay, T. K. (2014). A review on the modified finite point method. *Mathematical Problems in Engineering*.

Zhang. (2023). Application of finite element analysis in structural analysis and computer simulation. *Applied Mathematics and Nonlinear Sciences*, *9*.10.2478/amns.2023.1.00273

About the Contributors

Abdulsattar Abdullah Hamad was born in Salah Al-Deen, Iraq, He received Ph.D. from Madurai Kamaraj University-(India), School of Mathematics, and working at University of Samarra, His area of interest includes Dynamic Systems, Topology, Big Data, Machines Learning, Internet of Things, cloud computing, computational complexity, data analysis, geographic information systems, mobile ad hoc networks, power-aware computing, storage management, Served Guest editor in several journals of Inderscience, River Publishers, IGI, Journal of Information Science and Engineering and some of Scopus Journals, Also, have published research articles in Springer, IEEE Access, SCI, SCIE, Scopus indexed peer-review journals. And two book chapters in Springer, He get two PATENTS with Titles “Artificial Intelligence Based Automated Material Management System-Application No 202141024010, “Blockchain-Based Statistical Market Place Data Analysis System for Data Consumers through Fair Exchange- Application No 202141041466, He is serving as a reviewer in More than +160 Journals.

Sudan Jha is a Senior member of IEEE and a Professor in the Department of Computer Science & Engineering at Kathmandu University with over 22 years of combined teaching, research, and industrial experience. Nepal. His previous affiliations include KIIT University, Chandigarh University, Christ University, etc. He was 'Technical Director' at Nepal Television, 'Principal' at Nepal College of IT, and 'Individual Consultant' at Nepal Telecom Authority. Prof. Jha is dedicated to advancing higher education quality and actively works on smart platforms. His extensive research portfolio comprises 80+ published research papers and book chapters in esteemed SCI, SCIE indexed journals and conferences. He serves as an Editor-in-Chief of an international journal and Guest Editor for SCIE/ ESCI/SCOPUS indexed journals. With three patents to his name, he has authored and edited 6 books on cutting-edge topics in IoT, 5G, and AI, published by Elsevier, CRC, and AAP. His research has also secured funding for two international projects. Additionally, Prof. Jha contributes as a Keynote in more than 25 international conferences. In addition, he has also delivered faculty development programs, short-term training programs, workshops in national and international conferences, universities. He holds certifications in Microservices Architecture, Data Science, and Foundations of Artificial Intelligence. His primary research interests encompass Quality of Services in IoT enabled devices, Neutrosophic theory, and Neutrosophic Soft Set Systems.

***

Shihab A. Shawkat received his BSc in Computer Science from University of Tikrit in 2007 and MSc from Mansoura University in 2017. Currently, he is a PhD student in National School of Electronics and Communications of Sfax (ENET'Com), University of Sfax. He worked as a teacher during the period from 2008 to 2019 in Ministry of Education, Iraq. He has recently started working at the University of Samarra at the end of 2019 till now. His research interest lies in information security, image processing and AI.

Index

A

B

C

D

E

F

G

M

N

P

S

T

V

X